G. Meier · E. Sackmann · J. G. Grabmaier

Applications of Liquid Crystals

Springer Verlag

Berlin · Heidelberg · New York 1975

Dr. Gerhard Meier
Institut für Angewandte Festkörperphysik
der Fraunhofer-Gesellschaft, Freiburg im Breisgau

Professor Dr. Erich Sackmann
Abteilung für Experimentelle Physik III
der Universität Ulm

Dr. J. G. Grabmaier
Forschungslaboratorien der Siemens AG, München

ISBN-13:978-3-642-80956-9   e-ISBN-13:978-3-642-80954-5
DOI: 10.1007/978-3-642-80954-5

Library of Congress Cataloging in Publication Data. Meier, Gerhard, 1930. — Liquid crystals. Includes
bibliographies and index. CONTENTS: Meier, G. Physical properties of liquid crystals. — Sackmann, E.
Scientific applications of liquid crystals. — Grabmaier, J. G. Medical and technical applications of liquid
crystals. 1. Liquid crystals. 2. Liquid crystals devices. I. Sackmann, Erich, joint author. II. Grabmaier J. G.,
joint author. QD 923. M 44 548'.81 75-14496.

Typesetting: H. Charlesworth & Co LTD, Huddersfield.

Preface

Over the past ten years liquid crystals have attracted much interest and considerable progress has been made with respect to our knowledge in this field. The recent development was initiated mainly by the work of J. L. Fergason and G. H. Heilmeier, who pointed out the importance of liquid crystals for thermographic and electro-optic applications.

The first part of this book is a brief introduction to the physics of liquid crystals. The structures and properties of the three basic types of liquid crystals are discussed. A special paragraph is devoted to electric-field effects, which are important in display applications.

The chapter on Scientific Applications gives an insight into the potential applications of liquid crystals in fundamental research, with special emphasis on explaining the principles involved. Two groups of potential applications are discussed in detail:

1. the use of liquid crystals as anisotropic solvent for the determination of molecular properties by means of spectroscopy, and
2. their use in analytical chemistry, particularly in gas chromatography.

The reverse process involves the use of the dissolved molecules as microscopic probes in the investigation of the dynamical molecular structure of anisotropic fluid systems (e.g. biological membranes). This extremely important technique is also described.

The third part of the book is devoted to the medical and technical applications of liquid crystals. The colour of cholesteric liquid crystals changes with changes in temperature or other variables of state. This feature can be used in thermographic applications, for instance, thermal mapping of human skin for the diagnosis of circulatory diseases or the detection of tumors. In modern industrial applications this property is of value for direct temperature diagrams, detection of wave fields, locating faults in electronic devices and for thermally activated information displays.

When nematic liquid crystals are arranged in thin layers, their ability to transmit scattered or polarized light changes by applying an electric field. This phenomenon can be utilized for alphanumeric and analog displays, image converters, and matrix-type picture screens.

Today an indicator device is expected to assimilate the data rapidly and display them clearly, often in colour. It also should be compact, consume a minimum of electrical power, and require only low-level driving voltages. All these requirements are met by liquid crystal displays.

This book thus gives an overview of the theory and possible scientific, technical, and medical applications of liquid crystals. Its appeal is not only to physicists and chemists (especially spectroscopists) but equally to those in the manufacturing and processing industries (including electrical engineers).

Dr. J. G. Grabmaier
Dr. G. Meier
Prof. Dr. E. Sackmann

Contents

Physical Properties of Liquid Crystals

Gerhard Meier
Institut für Angewandte Festkörperphysik der Fraunhofer-Gesellschaft
D-7800 Freiburg i. Br., Federal Republic of Germany

1. Introduction

The term "Liquid Crystals", first used by O. Lehmann in 1890, designates a state of matter that is intermediate between the solid crystalline and the ordinary (isotropic) liquid phases. Liquid crystals flow like ordinary liquids, e.g. they adopt the shape of their container. On the other hand, they exhibit anisotropic properties as do solid crystals. One can therefore define liquid crystals as "condensed fluid phases with spontaneous anisotropy". Liquid crystals are also called mesophases or mesomorphic phases because of their intermediate nature.

Many thousands of organic compounds form liquid crystals when the solid crystal is heated above its melting point. The mesomorphic phase appears as a more or less viscous fluid which can be identified visually by its characteristic turbidity or with a polarizing microscope by its optical birefringence. At higher temperatures, transitions to other mesophases may occur in some cases, while other compounds display only one mesophase. In either case, at another well defined higher transition ·

temperature, the turbidity suddenly vanishes giving way to the clear appearance of the ordinary liquid. This phase transition is reversible and of the first order type with a latent heat of some 100 cal/mole.

Such liquid crystals which are formed when the temperature is varied are called thermotropic. Many compounds form a second kind of mesophase when a solvent is added. Further increasing the amount of solvent beyond a critical concentration leads to a transformation to an ordinary (isotropic) liquid. Mesomorphic phases of this kind are called lyotropic. They are formed e.g. with long-chain saturated fatty acids in aqueous solutions. Lyotropic liquid crystals play an important role in biological systems. However, no technical applications are known as yet, and therefore we will not discuss this type of mesophases further.

In the following paragraphs, we shall introduce some aspects of liquid crystals which are important for applications. The reader who is interested in more fundamental details is referred to the various review articles [1−7] and books [8−10] on this subject.

2. Basic Types and Structures of Liquid Crystals

2.1. Classification after Friedel

There are three basic types of liquid crystals. Following a proposal by Friedel [11] they are denominated as smectic, nematic and cholesteric. A schematic representation of these three basic types is given in Fig. 1.

Many compounds do not form only one of these types, but exhibit above the melting point first one or more smectic types and at higher temperatures either the nematic or the cholesteric type. The cholesteric mesophase is formed by compounds or mixtures the molecules of which are chiralic (non-superimposable with their mirror image). The various types of mesophases differ in the orientational order of the long molecular axes. In all cases known as yet, this ordering is apolar and no ferroelectric behavior has been observed.

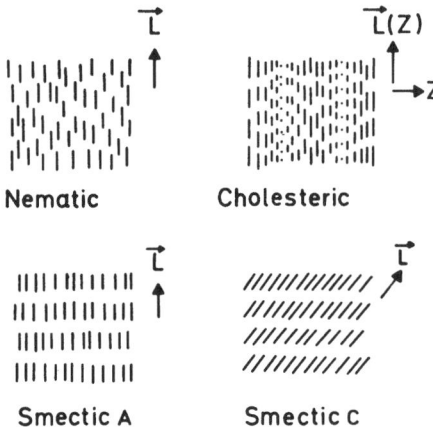

Figure 1. Structures of liquid crystals.

2.2. Smectic Liquid Crystals

Friedel's classification had to be extended to account for the fact that there are
various types of smectic liquid crystals denominated by A, B, C, ... All smectic
liquid crystals have a layered structure. The centers of gravity of the elongated
molecules are arranged in equidistant planes. The long axes of the molecules are
parallel to a preferred direction $\vec{L}$ which may be normal to the planes (smectic A)
or tilted by a certain angle (smectic C). The arrangement of the centers of gravity
within the planes may be at random or regular.

Information about the molecular ordering has been obtained from X-ray
analysis. The various smectic types can be distinguished by their textures in a
polarizing microscope. According to Demus et al. [12] at least seven different
types have been observed. A typical example for a compound forming smectic
phases is p-(p'-ethoxybenzylideneamino)-ethylcinnamate

$$C_2H_5O-\bigcirc-CH=N-\bigcirc-CH=CH-COOC_2H_5$$

with the phase sequence solid – (81°C) – smectic B – (118.5°C) – smectic A –
(156.5°C) – nematic – (159°C) – isotropic.

Up to now, smectic liquid crystals have obtained only little importance for
applications and therefore this type will be discussed no further.

2.3. Nematic Liquid Crystals

Nematic liquid crystals differ from ordinary liquids by a long range orientational
order of the long molecular axes which is characterized by one single principle:
averaged over time or space, the long molecular axes of the molecules are aligned
parallel to a preferred direction $\vec{L}$. As long as no ferroelectric effects can be
detected the alignment must be assumed to be nonpolar. The centers of gravity
are distributed at random as in an ordinary liquid (we neglect the small short range
order occurring in all fluid systems). The molecules are allowed to rotate freely
about their long axes. Following from this structure, liquid crystals are uniaxial
with respect to all physical properties. The axis of symmetry is identical with the
preferred axis of the structure, $\vec{L}$. According to the molecular statistical theory by
Maier and Saupe [13], the anisotropy of the molecular polarizability is responsible
for the occurrence of the liquid crystalline phase. The anisotropy of the polarizability, in turn, is to a first approximation determined by the geometrical anisotropy.
Therefore, mesomorphic behavior is found in compounds with elongated molecules
which in most cases consist of a conjugated aromatic system. The relation between
molecular structure and mesomorphic behavior has been discussed in more details
by Gray and Winsor [8] and de Jeu and van der Veen [14].

Many compounds forming nematic liquid crystals have the simple structure

$$R-\bigcirc-X-\bigcirc-R'$$

The most important groups X linking the two substituted phenyl rings are listed in Table 1. The reader interested in the methods of chemical synthesis is referred to the literature given there. R and R' are small groups or rather short chains. Examples for compounds that exhibit a nematic phase are given in Table 2. The best known compound is p-azoxyanisole (1). A very important room temperature nematic liquid crystal is MBBA (2 in Table 2). An extension of the liquid crystalline temperature range can be achieved by mixing MBBA with EBBA (3). Other low melting materials are the azoxy compounds (4), the phenyl benzoates

Table 1.

– X –	Name	Ref.
– N = N –	azobenzenes	[15, 16]
– N = NO –	azoxybenzenes	[17, 18]
– CH = CH –	stilbenes (trans)	[19]
– C ≡ C –	tolanes	[20]
– CH = N (O) –	nitrones	[21]
– CH = N –	Schiff bases	[22, 23]
– O – CO –	phenyl benzoates	[24]
–	biphenyls	[25]

Table 2.

No.	Name	Formula	Nematic Range °C
(1)	p-azoxyanisole (PAA)	H_3CO—⬡—N=N—⬡—OCH_3 (O)	118 to 136
(2)	4-methoxy-4'-n-butyl-benzylideneaniline (MBBA)	H_3CO—⬡—CH=N—⬡—C_4H_9–n	21 to 48
(3)	4-ethoxy-4'-n-butyl-benzylideneaniline (EBBA)	H_5C_2O—⬡—CH=N—⬡—C_4H_9–n	37 to 80
(4)	4-methoxy-4'-n-butyl-azoxybenzene	H_3CO—⬡—N=N—⬡—C_4H_9–n (O) H_3CO—⬡—N=N—⬡—C_4H_9–n (O)	16 to 76
(5)	p-hexylcarbonato-p'-heptoxyphenyl benzoate	$H_{13}C_6OCOO$—⬡—COO—⬡—OC_7H_{15}	36 to 54
(6)	(4'-ethoxybenzylidene)-4-aminobenzonitrile (PEBAB)	H_5C_2O—⬡—CH=N—⬡—C≡N	106 to 128
(7)	4'-cyanophenyl-4-n heptylbenzoate 4'-cyanophenyl-4-n-butylbenzoate	$H_{15}C_7$—⬡—COO—⬡—C≡N H_9C_4—⬡—COO—⬡—C≡N	25 to 50
(8)	4'-n-pentyl-4-cyanobiphenyl (PCB)	$H_{11}C_5$—⬡—⬡—C≡N	22,5 to 35

(5, 7 and 8), and the cyano biphenyls (8). A recent review on nematic liquid crystalline materials that are useful for electro-optical application has been given by Castellano (26).

2.4. Cholesteric Liquid Crystals

Cholesteric liquid crystals are denominated after cholesterol since many cholesterol esters form this type of liquid crystals (cholesterol itself exhibits no mesomorphic phase). The structure can be described as a twisted nematic structure: considering a certain plane, the molecules are aligned parallel to a preferred direction $\vec{L}$ as in a nematic phase. When proceeding in a direction normal to the plane, $\vec{L}$ rotates continuously. The result is a helical structure, the axis of which can be described by a unit vector $\vec{Z}$. The pitch p of the structure is defined by the distance which is necessary to rotate $\vec{L}$ by 2π when proceeding along $\vec{Z}$. p varies widely and falls between 0.2 μm and macroscopic values. An infinite pitch corresponds to the nematic structure. The pitch can be adjusted to any desired value by adding appropriate amounts of a chiralic compound to a nematic liquid crystal. The sign of rotation of the helix may be left-handed (type "laevo") or right-handed (type "dextro"). Most cholesterol derivatives form type laevo, but cholesterol chloride is an example for type dextro. There is no correlation between the sign of the optical activity of the free molecule (i.e. in dilute solution) and the type of the cholesteric helix. However, a necessary condition for the formation of the cholesteric mesophase is that the molecules are chiralic.

3. Physical Properties and Theories of Nematic Liquid Crystals

3.1. Alignment of Nematic Liquid Crystals

In a macroscopic volume of a nematic liquid crystal, the preferential direction $\vec{L}$ is in general not uniform, but changes from place to place under the action of disturbing forces such as convection flow, wall effects, etc. However, even large samples can be aligned by both magnetic and electric fields. Contrary to the very weak field effects in ordinary liquids, this field-induced alignment of a liquid crystal is a strong effect. This is due to the peculiar intermolecular forces responsible for the mesomorphic state and lead to a cooperative behavior of the molecules. The alignment depends on the sign of the susceptibility anisotropy $\Delta\chi = \chi_\parallel - \chi_\perp$, where $\chi_\parallel$ and $\chi_\perp$ are the electric or magnetic susceptibilities parallel and perpendicular to the structural symmetry axis $\vec{L}$. For $\Delta\chi$ being positive, $\vec{L}$ is aligned parallel to the applied field. The dielectric anisotropy, $\Delta\chi_e$, may be positive or negative (see 3.5). The magnetic susceptibilities $\chi_\parallel$ and $\chi_\perp$ are negative (diamagnetic materials) and very small (10^{-6} to 10^{-7}). $|\chi_\perp|$ is larger than $|\chi_\parallel|$ for usual pure nematic systems so that $\Delta\chi$ becomes positive and $\vec{L}$ is aligned parallel to an applied magnetic field. However, negative values of $\Delta\chi$ have been observed with compensated nematic mixtures consisting of cholesteryl derivatives and with certain lyotropic liquid crystals (Part II of this book).

Thin films of liquid crystals (up to 100 μm) can be aligned by surface-generated orienting forces which result from physico-chemical processes, e.g. hydrogen bonding, Van der Waal's interactions, or dipolar interactions, and from mechanical interactions as a result of the anisotropic elasticity of the liquid crystal. For flat, untextured surfaces, the physico-chemical interactions prevail while for textured surfaces the anisotropic elastic interactions must also be considered. Depending on the nature of substrates and surface treatment, the symmetry axis $\vec{L}$ can be aligned parallel or normal to the plane of the layer [27–30]. Uniformly oriented liquid crystal layers are required for most liquid crystal device applications. Several methods of chemical and mechanical surface treatment are known which will be treated on pages 27–28 and 97–102.

3.2. Order Parameter

In a homogeneously ordered liquid crystal sample, the long axes of the molecules are all parallel to a preferred direction which is represented by the unit vector $\vec{L}$. This holds only in a time or space average. In reality, due to their thermal energy the individual molecules or small groups of molecules tumble about the preferred direction. The efficiency of the molecular orientation along $\vec{L}$ can be described by a single order parameter S:

$$S = \frac{1}{2} < 3 \cos^2 \theta - 1 >,$$

where the brackets mean the time or space average and θ is the angle between the long molecular axis and the preferred direction (optic axis). S = 1 means that all molecular axes are parallel. This ideal nematic order would be possible near the absolute zero point of temperature only if the material would not freeze. For an isotropic liquid S = 0. Real nematic liquid crystals exhibit order parameters about 0.4 to 0.7, depending on temperature. The value of S is not substantially changed by applied magnetic or electric fields of usual laboratory magnitude.

The order parameter S can be determined by measuring the principal refractive indices or the diamagnetic susceptibilities. Other methods use the ultraviolet or infrared dichroism, nuclear magnetic resonance spectra, or X-ray diffraction. A critical comparison of the various methods was given by Saupe and Maier [31].

3.3. Molecular Statistical Theory

Maier and Saupe [13] developed a statistical theory to describe the liquid crystalline state and the molecular ordering for the nematic phase. In analogy to the treatment of ordering phenomena in ferromagnetics or ferroelectrics, this theory describes the intermolecular orientational forces by a mean field method. Each individual molecule feels a "nematic potential" D = f (θ, S, V) which depends on the momentaneous angle θ between its long axis and the optic axis, the order parameter S and the molar volume V. S is then given by

$$S = 1 - \frac{3}{2} \int_0^{\pi/2} \exp\left\{-\frac{D}{kT}\right\} \sin^3 \theta \, d\theta \bigg/ \int_0^{\pi/2} \exp\left\{-\frac{D}{kT}\right\} \sin \theta \, d\theta$$

as a function of temperature when D is known. In the Maier—Saupe theory D is given by

$$D = -\frac{A}{V^2} S \left(1 - \frac{3}{2} \sin^2 \theta\right).$$

A is a quantity which is characteristic for each compound and can be calculated from the transition temperature and molar volumes just below and above the transition to the isotropic phase. The derivation is based on the model that the molecular alignment is caused by dispersion forces and only the induced dipole—dipole contribution is considered. The anisotropy of the molecular polarizability causes an angular dependence of the intermolecular dispersion forces and therefore is responsible for the mesomorphic state.

3.4. Elastic Continuum Theory

Liquid crystals exhibit elastic properties which can be described by the continuum theory proposed by Zocher [32] and Oseen [33]. Frank [34] presented this elastic theory in a more simple and comprehensive form.

The elastic continuum theory is based on the assumption that at each point within the liquid crystal a preferential direction for the molecular orientation is given which is described by a unit vector $\vec{L}$, and which varies continuously from place to place — except for a few singular lines or points. Any distortion of the undisturbed state requires a certain amount of energy since elastic torques attempt to maintain the original configuration. The elastic energy density of a deformed nematic liquid crystal is given by

$$g_E = \frac{1}{2} k_{11} (\text{div } \vec{L})^2 + \frac{1}{2} k_{22} (\vec{L} \text{ curl } \vec{L})^2 + \frac{1}{2} k_{33} (\vec{L} \times \text{curl } \vec{L})^2$$

k_{11}, k_{22} and k_{33} are the elastic curvature constants of splay, twist and bend respectively. The elastic constants are of the order 10^{-5} to 10^{-6} dyn. The three fundamental deformations of a nematic liquid crystal are schematically represented in Fig. 2.

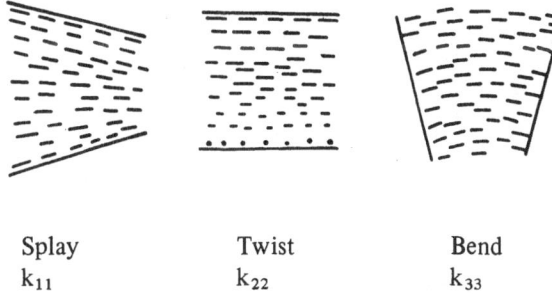

Splay   Twist   Bend
k_{11} k_{22} k_{33}

Figure 2. Basic deformations of a nematic liquid crystal.

3.5. Dielectric Properties

The dielectric behavior of a nematic liquid crystal is described by two dielectric constants: $\epsilon_\parallel$ and $\epsilon_\perp$, as measured parallel and perpendicular respectively to the axis of symmetry. The dielectric anisotropy, $\Delta\epsilon = \epsilon_\parallel - \epsilon_\perp$, plays an important role in all electrooptic applications of liquid crystals. $\Delta\epsilon$ may be positive or negative [35]. PAA, e.g., has a negative anisotropy (Fig. 3). In Table 2, $\Delta\epsilon$ is negative for compounds (1) through (5) and positive for compounds (6) through (8).

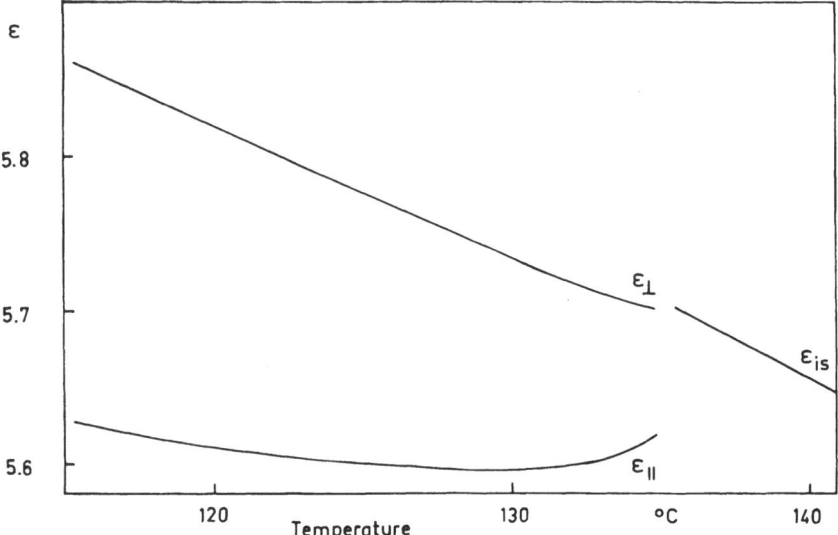

Figure 3. Dielectric constants of p-azoxyanisole: $\epsilon_\parallel$ and $\epsilon_\perp$ in the nematic phase, ϵ_{is} in the isotopic phase (after Maier and Meier [35]).

The sign of the dielectric anisotropy depends on the molecular polarizabilities and the value and angular position of the permanent electric dipole moment. In the case of p-azoxyanisole, there is a strong dipole moment that makes a large angle with the long axis of the molecule. Therefore the orientational polarization is much larger perpendicular to $\vec{L}$ than parallel to $\vec{L}$, and $\epsilon_\perp$ is larger than $\epsilon_\parallel$. Nematic liquid crystals composed of molecules with the strongly polar $-C \equiv N$ group in the terminal position have large positive dielectric anisotropies in the range of 10 to 20 [25, 36, 37, 38].

Maier and Meier [35] worked out these ideas more quantitatively by applying Onsager's theory of static polarization for nematic liquid crystals. The Onsager theory relates the dielectric constant to molecular properties, namely the molecular polarizability α and the permanent electrical dipole moment μ. In the case of nematic liquid crystals, however, the polarizability has to be treated as a tensor with principal values $\alpha_\parallel$ and $\alpha_\perp$ and an angle β has to be introduced to describe the position of the permanent dipole moment with respect to the long axis of the molecule. Then follows

$$\epsilon_\parallel = 1 + 4\pi NhF \left\{ \bar{\alpha} + \frac{2}{3}\Delta\alpha S + F\frac{\mu^2}{3kT}[1 - (1 - 3\cos^2\beta)S] \right\}$$

$$\epsilon_\perp = 1 + 4\pi NhF \left\{ \bar{\alpha} - \frac{1}{3}\Delta\alpha S + F\frac{\mu^2}{3kT}[1 + \frac{1}{2}(1 - 3\cos^2\beta)S] \right\}$$

The factors h and F are given by

$$h = \frac{3\bar{\epsilon}}{2\bar{\epsilon} + 1}; \quad F = \frac{1}{1 - \bar{\alpha}f}; \quad f = \frac{2\bar{\epsilon} - 2}{2\bar{\epsilon} + 1}\frac{4}{3}\pi N; \quad \bar{\epsilon} = \frac{\epsilon_\parallel + 2\epsilon_\perp}{3}$$

N is the number of molecules per cm³; $\bar{\alpha} = (\alpha_\parallel + 2\alpha_\perp)/3$; $\Delta\alpha = \alpha_\parallel - \alpha_\perp$; S is the order parameter. These formulae were derived assuming that the intrinsic field is isotropic, which is not exactly true but should be a good approximation if the dielectric anisotropy is not too large. The dielectric anisotropy is given by

$$\Delta\epsilon = 4\pi NhF \left[\Delta\alpha - F\frac{\mu^2}{3kT} \cdot \frac{3}{2}(1 - 3\cos^2\beta) \right] S.$$

The first term represents the contribution of the polarizability (displacement polarization) and is always positive whereas the second term originates from the permanent dipole moment (orientation polarization) and is positive or negative depending on whether β is smaller or larger than 55°.

In nematic liquid crystals, $\epsilon_\parallel$ and $\epsilon_\perp$ exhibit a normal Debye relaxation at microwave frequencies. However, in some compounds $\epsilon_\parallel$ has an additional region of relaxation at much lower frequencies, whereas $\epsilon_\perp$ shows no dispersion in this region. The low frequency relaxation process can be characterized by a relaxation time τ_1 which is longer than the ordinary Debye relaxation time τ_0 by a "retardation factor" $g = \tau_1/\tau_0$. g has an order of magnitude of 100 to 1000 and is strongly temperature-dependent. This low frequency relaxation is assigned to the permanent dipole component parallel to the long molecular axis. This component can contribute to $\epsilon_\parallel$ only by a rotation of the molecule about a transverse axis. For this to happen it is necessary to overcome the barrier of the potential which is present in the mesomorphic state and which tries to align all molecules parallel to each other. It is possible to extend the Debye theory of dipole relaxation to include the aligning action of this "nematic potential" and to show how the retardation factor g depends on the nematic order [39].

4. Optical Properties of Cholesteric Liquid Crystals

Because of their helical structure, cholesteric liquid crystals exhibit unique optical properties when they are in the planar texture, e.g. when the liquid crystal layer is uniformly aligned in such a way that the helical axis $\vec{Z}$ is normal to the substrate. The most striking feature of such a uniformly aligned cholesteric liquid crystal is the selective light reflection (Fig. 4) which is the origin of the brilliant colors that are displayed under certain conditions: in a small wavelength range about λ_0 a light

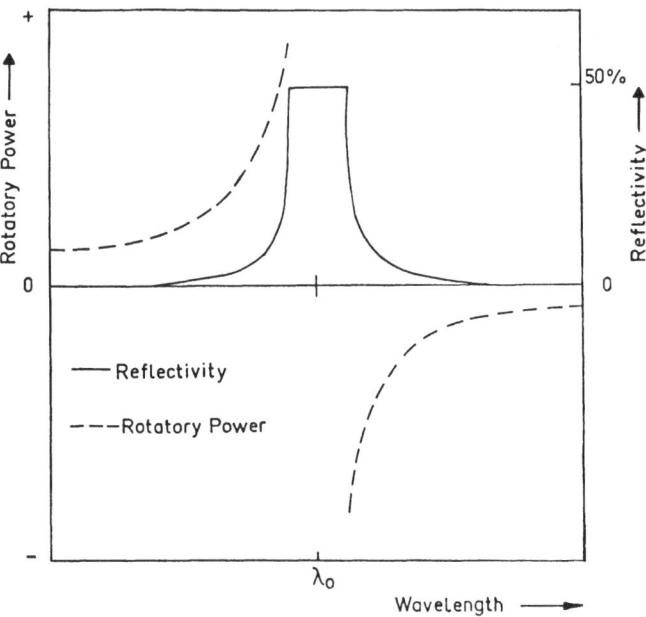

Figure 4. Optical properties of a cholesteric liquid crystal (Type "dextro"; incident light linearly polarized)

beam incident parallel to the helical axis is split into its two circularly polarized components, one of which is transmitted whereas the other one is totally reflected. If one defines the sense of rotation of the light vector looking against the beam, the rotation of the reflected circularly polarized light agrees with the screw sense of the structure. Maximum reflection occurs when

$$\lambda_0 = \bar{n} \cdot p.$$

$\bar{n} = (n_{\parallel} + n_{\perp})/2$ is the mean refractive index within a plane normal to $\vec{Z}$. The width $\Delta\lambda$ of the reflection band is given by

$$\Delta\lambda = \Delta n \cdot p,$$

where $\Delta n = n_{\parallel} - n_{\perp}$ is the anisotropy of the refractive index. For obliquely incident light the reflection band is shifted to shorter wavelengths and higher order reflection bands occur which are absent at normal incidence. Theoretical treatments of the selective reflection of cholesteric liquid crystals were given by several authors [40–44].

Outside the reflection band, cholesteric liquid crystals exhibit an extremely strong optical rotatory power which can amount to values as large as $20.000°$ or 50 complete rotations per millimeter. The sign of the rotatory power is different below and above the reflection band.

In many compounds, the pitch p and hence the reflection color is strongly temperature-dependent. Thin homogeneously aligned layers of such compounds therefore can be used to display directly the temperature distribution of the substrate (see Part III of this book). Usually the pitch decreases and the reflection band is shifted to shorter wavelengths when the temperature is increased, but other cases with a reverse temperature dependence are known. The temperature sensitivity becomes extremely strong in the neighborhood of a cholesteric to smectic phase transition temperature (Fig. 5).

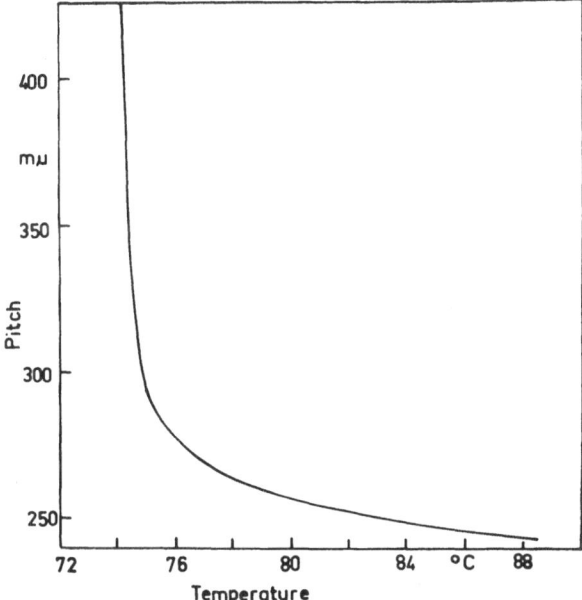

Figure 5. Pitch of the helix as a function of temperature for cholesteryl nonanoate (after Kassubek and Meier [45])

5. Electric Field Effects in Nematic Liquid Crystals

5.1. Influence of Electric Fields on the Structure

External fields of usual laboratory size do not noticeably change the order parameter of a liquid crystal. However, they do influence the macroscopic structure of the liquid crystal, which in turn influences the optical properties. Various electro-optic effects are known which can be directly applied in display devices.

Electric effects in liquid crystals are usually studied in cells with a sandwich structure: the liquid crystal material is contained between two closely spaced (6 μm to 25 μm) plates, both coated on the inside with a transparent conductivity layer. The optic axis of the liquid crystal is uniformly aligned by special treatment of the electrode surfaces.

Three groups of electrically induced effects are to be distinguished: (a) Dielectric effects. These are analogous to magnetically induced deformations and are purely field induced, i.e. they need no electric current. In the equilibrium state, no fluid flow is present. Dielectric effects are important for a number of display applications. (b) Piezoelectric or flexoelectric effects. Meyer [46] has shown that bend or splay deformations of nematic liquid crystals result in an electric polarization if the molecules have a certain shape anisotropy. This "piezoelectric effect" (Meyer) or "flexoelectric effect" (de Gennes) has been investigated by several authors [47, 48], but no practical applications are known as yet. Therefore this kind of effect will not be treated further here. (c) Electrohydrodynamic effects. These occur in materials with negative dielectric anisotropy when the electric conductivity is not too small and/or the frequency of the applied field is not too high. They are connected with an electric current and a fluid flow. One example is the dynamic scattering effect which is very important for display applications.

5.2. Dielectric Effects

Three important dielectric effects in nematic liquid crystals are to be discussed. The three cases differ in the geometry of the specimen and the sign of dielectric anisotropy of the material used (Fig. 6a–c).

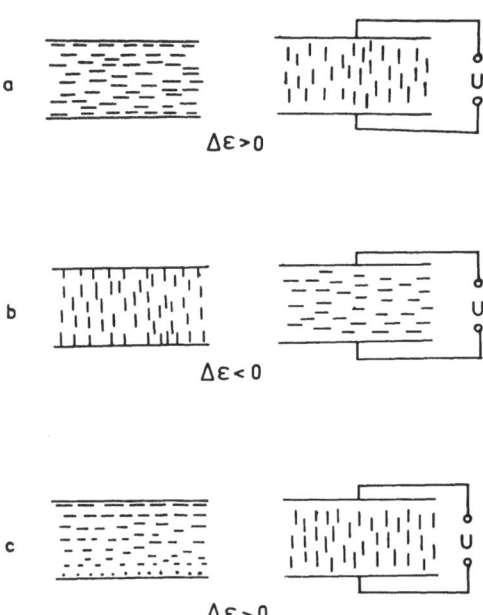

Figure 6. Dielectric deformations in a nematic liquid crystal
(a) Parallel orientation: Fréedericksz effect
(b) Perpendicular orientation: DAP effect (electrically controlled birefringence)
(c) Twist cell: Schadt-Helfrich effect

5.2.1. Parallel Orientation: Fréederisksz-Effect

Consider a configuration where the undisturbed optic axis of the liquid crystal is aligned parallel to the surface of the electrodes. The dielectric anisotropy $\Delta\epsilon$ is assumed to be positive. The analogous magnetic case was treated by Saupe [49]. The problem of the deformation by an electric field, which is more difficult because of the large dielectric anisotropy, was solved by several authors [50, 51].

Under the influence of an electric field two competitive actions are to be taken into account: because of the positive dielectric anisotropy the electric field exerts a torque on each volume element trying to align the optic axis along the field and perpendicular to the electrodes. On the other hand, the interaction with the boundary surfaces causes elastic torques which try to restore the undisturbed state. Below a critical threshold voltage U_o, the restoring elastic torque is larger than the electric torque so that no deformation occurs at all. Only above U_o does a deformation take place. U_o is given by

$$U_o = \pi \sqrt{\frac{k_{11}}{\epsilon_o \cdot \Delta\epsilon}}$$

k_{11} is the elastic constant for splay deformation. ϵ_o is a dimensional constant. U_o usually falls between 3 V and 6 V.

When a voltage $U > U_o$ is applied, the angle between the local optic axis and the applied field is a function of U and z, z being the coordinate normal to the electrodes. For U large enough, the optic axis is uniformly aligned normal to the electrodes: homeotropic alignment. Between crossed polarizers, the homeotropically aligned cell appears dark. The transient response behavior of the liquid crystalline structure depends on the elastic and dielectric properties of the material, on the applied voltage and on the thickness of the sample used. The processes of deforming and relaxing are connected with a fluid flow, and therefore also viscous forces come into play. The mathematical treatment of the problem is extremely difficult. However, reasonable simplifications, e.g. small angle approximation, lead to simple expressions for rise and decay times:

$$\tau_{rise} = \frac{d^2}{\pi^2} \frac{\eta_1}{k_{11}} \left[\left(\frac{U}{U_o}\right)^2 - 1 \right]^{-1}$$

$$\tau_{decay} = \frac{d^2}{\pi^2} \frac{\eta_1}{k_{11}},$$

where d is the thickness of the layer and η_1 a viscosity parameter. The dielectric anisotropy, $\Delta\epsilon$, enters by the threshold voltage, U_o.

Rise and decay times are proportional to the square of the thickness. The rise time additionally depends on the applied voltage according to the factor $[(U/U_o)^2 - 1]^{-1}$. For a layer thickness of 10 μm to 20 μm typical values fall between 10 ms and 30 ms for the rise time and 30 ms to 100 ms for the decay time.

5.2.2. Perpendicular Orientation: DAP – Effect (Electrically Controlled Birefringence)

A similar effect occurs in a configuration where the optic axis of the liquid crystal is aligned normal to the electrodes (homeotropic alignment) when the dielectric anisotropy $\Delta\epsilon$ is negative. The electric conductivity of the material is assumed to be very small so that electrohydrodynamic effects are avoided. In this case, the threshold voltage is given by

$$U_o = \pi \sqrt{\frac{k_{33}}{\epsilon_o(-\Delta\epsilon)}};$$

k_{33} is the elastic constant for bend deformation. U_o amounts to 3V to 6V for most materials.

When a voltage $U > U_o$ is applied, the sample is deformed. The retardation between the wavefronts of the ordinary and the extraordinary rays of transmitted light is dependent on the applied voltage. The effect is called "DAP-effect" (Deformation of Aligned Phases) or "electrically controlled birefringence" and can be applied in display devices [52–55] (see third part of this book).

The transient response times are given by

$$\tau_{\text{rise}} = \frac{d^2}{\pi^2} \frac{\eta_2}{k_{33}} \left[\left(\frac{U}{U_o}\right)^2 - 1 \right]^{-1},$$

$$\tau_{\text{decay}} = \frac{d^2}{\pi^2} \frac{\eta_2}{k_{33}},$$

where η_2 is a viscosity parameter. Typical numerical values are of the same order of magnitude as in the case of parallel orientation.

5.2.3. Twist-cell: Schadt–Helfrich-Effect

A new electro-optic effect was discovered by Schadt and Helfrich [56]. The effect occurs in a nematic liquid crystal having a structure that has been continuously twisted by $90°$ by treating the surfaces of the electrodes such that the optic axis of the liquid crystal is aligned parallel to the electrodes, the preferred directions on the two electrodes being at right angles to each other. The nematic liquid crystal has a positive dielectric anisotropy. The polarization vector of light transmitted through this cell follows the twisted structure. When an electric field is applied a deformation occurs above a threshold voltage U_o given by

$$U_o = \pi \sqrt{\frac{k_{11} + (k_{33} - 2k_{22})/4}{\epsilon_o \Delta\epsilon}}.$$

For a voltage U large enough the optic axis reorients to be normal to the electrodes. In this state the polarization vector of the transmitted light is not affected. The effect has a threshold voltage U_o of about 3 V to 5 V, but values of as low as 1 V have been reported for special compounds [25, 37, 38].

In display applications, the twist cell is placed between two polarizers. They can either be oriented parallel or perpendicular to each other, the switching modes of these two configurations being complementary. The electro-optic characteristics of nematic twist cells have been studied by several authors [56–59]. Applications will be treated in Part III of this book.

5.3. Electrohydrodynamic Effects

Electrohydrodynamic effects in liquid crystals are connected with an electrical current and a mechanical fluid flow. Consequently, the anisotropic electric conductivity and the various viscosity parameters play an important role.

Several electrohydrodynamic effects are known. The most important effect is dynamic scattering as described in 1968 by the RCA liquid crystal group [60]. Already in 1918 Björnstahl [61] reported measurements on the extinction of light transmitted through a nematic liquid crystal under the influence of an electric field. With a 6.2 mm thick layer of p-azoxyphenetole, he observed a threshold field of 100 V/cm, above which the extinction increased sharply.

Dynamic scattering can easily be observed in a transparent sandwich cell filled with a nematic liquid crystal which has a negative dielectric anisotropy. When an increasing voltage is applied, a striped pattern appears, which is called Williams domains [62]. On further increasing the voltage the regularly shaped pattern is destroyed subsequently giving way to a turbulent state which strongly scatters light ("dynamic scattering").

When the surfaces of the electrodes have been treated to align the undisturbed optic axis of the nematic liquid crystal uniformly parallel to the electrodes, the Williams domain stripes are perpendicular to this original direction. By viewing with polarized light one can see that the optic axis in the deformed layer lies in the plane defined by the original alignment and the electric field vector. The periodicity of the pattern is approximately equal to the layer thickness and almost voltage-independent. The pattern appears only above a certain threshold voltage of about 3 V to 7 V. The effect is associated with a fluid flow as can easily be seen from the motion of suspended dust particles.

The interpretation of the formation of Williams domains has been given by Carr [63] and Helfrich [64]. According to this model, the anisotropies of the dielectric constant and of the electric conductivity play an important role. When a small disturbance of the originally parallel alignment is present then the anisotropy of the conductivity leads to the formation of space charges. The applied electric field exerts forces on the charged volume elements which try to increase the deformation. On the other hand, the original alignment is stabilized by elastic torques caused by the surface alignment and dielectric torques if the dielectric anisotropy is negative. At a certain threshold voltage, the deforming forces overcome the stabilizing forces and a deformation takes place.

The formation of space charges requires a certain time. We define a space charge relaxation time τ as the time which is necessary to allow the material to reach the electrostatic equilibrium after applying an electric field. τ is given by

$$\tau = \frac{\epsilon_0 \bar{\epsilon}}{2\pi\bar{\sigma}}.$$

For our purposes, the assumption of a single relaxation time τ described in terms of the average values $\bar{\epsilon}$ and $\bar{\sigma}$ is sufficient. $\bar{\sigma}$ is the electric conductivity.

When a.c. fields are used to study Williams domains or dynamic scattering, the frequency must be lower than the limiting frequency $f_0 = (2\pi\tau)^{-1}$ in order to allow space charge formation. f_o is the space-charge limited dielectric relaxation frequency.

The mathematical treatment of the Carr-Helfrich model leads to the following expression for the threshold voltage U_o for Williams domains [64]:

$$U_o = 2\pi \frac{d}{\lambda} \left\{ \frac{k_{33}}{\epsilon_o \epsilon_\parallel \left(\frac{\Delta\epsilon}{\epsilon_\parallel} \frac{\sigma_\perp}{\sigma_\parallel} - \eta' \left[\frac{\epsilon_\perp}{\epsilon_\parallel} - \frac{\sigma_\perp}{\sigma_\parallel} \right] \right)} \right\}^{\frac{1}{2}}.$$

d is the sample thickness, λ the period of the pattern, k_{33} the bend elastic constant and η' an effective viscosity which can be expressed by the viscosity parameters α_i (i = 1, 2, 3, 4, 5) defined by Leslie [65]:

$$\eta' = 2\alpha_2/(\alpha_4 - \alpha_2 + \alpha_5).$$

$\sigma_\parallel$ and $\sigma_\perp$ are the electric conductivities as measured parallel and perpendicular respectively to the axis of symmetry. In most nematic liquid crystals the parallel conductivity, $\sigma_\parallel$, is larger than the perpendicular conductivity, $\sigma_\perp$.

The period λ does not follow from the Carr-Helfrich theory. Experimental observations show that λ is nearly equal to the sample thickness d. Thus U_o is approximately independent from the sample thickness: the threshold is given by a voltage and not by a field strength.

The mathematical treatment of the dynamics of the Williams domains is a difficult problem. However, reasonable assumptions (small angle approximation, exponential time response to a stepwise change in voltage) lead to expressions for the time constants which are in qualitative agreement with experiment:

$$\tau_{rise} = \frac{d^2 \eta''}{4\pi^2 k} \left[\left(\frac{U}{U_o} \right)^2 - 1 \right]^{-1},$$

$$\tau_{decay} = \frac{d^2 \eta''}{4\pi^2 k},$$

where η'' is a viscosity parameter and k is the appropriate elastic constant.

So, rise and decay times are proportional to the square of the thickness, and the rise time additionally depends on the applied voltage according to the factor $[(U/U_o)^2 - 1]^{-1}$. For a layer thickness of 10 μm to 20 μm typical values fall between 10 ms and 50 ms for the rise time and 30 ms to 150 ms for the decay time.

In very thin liquid crystal layers Greubel and Wolff [66] observed a regular domain pattern having a voltage-dependent period and which can be used as a tunable diffraction grating providing an electronic color control. These domains clearly differ from the ordinary Williams domains. They occur in very thin samples (d $\leq$ 6 μm) having a rather low conductivity.

Williams domains and dynamic scattering mode represent states of liquid crystals in which the optical homogeneity is strongly disturbed. Especially in the dynamic scattering mode the nematic order is maintained only within small areas of a few micrometers in diameter which act as scattering centers for light. Transmitted light is strongly scattered because the effective refractive index varies rapidly on its path. This effect can be used in display devices. In addition, polarized light is depolarized after transmittance through a dynamic scattering cell. Applications will be described in Part III of this book.

6. Electric Field Effects in Cholesteric Liquid Crystals

6.1. Cholesteric-Nematic Phase Transition (Helical Unwinding)

The so-called phase change effect occurs in cholesteric liquid crystals if the dielectric anisotropy is positive and an electric field is applied at right angles to the helicoidal axis. The frequency should be high enough to avoid hydrodynamic effects. With increasing voltage the pitch increases and above a critical field the distorted structure is identical with a uniformly aligned nematic liquid crystal. This critical field E_c is given by [67, 68]:

$$E_c = \frac{\pi^2}{p_0} \sqrt{\frac{k_{22}}{\epsilon_0 \Delta \epsilon}},$$

where p_0 is the undisturbed pitch and k_{22} the twist elastic constant. Typical values of E_c are of the order of 10 V/μm.

The effect is purely field-induced. Even in the periods of turn-on and turn-off, the effect is not connected with a material flow. Therefore the response times are remarkably short. Rise and decay times as short as 10 μs have been observed [69].

6.2. Conical Deformation

Let us assume a planar texture (helical axis at right angles to the boundaries) and a positive dielectric anisotropy. When an electric field is applied parallel to the helical axis the local optical axis experiences a dielectric torque which tries to tip it out of the cholesteric plane. The boundary forces, on the other hand, act in a direction to sustain the planar texture. Assuming that the deformation is uniform in the plane of the sample, Leslie [70] was able to show that below a certain threshold voltage U_c no deformation occurs. Above U_c the local optic axis tips out of the cholesteric plane: "conical deformation". U_c is given by

$$U_c = \pi \sqrt{\frac{k_{11} + 4k_{33}(d/p_0)^2}{\epsilon_0 \Delta \epsilon}},$$

where k_{11} and k_{33} are the elastic splay and bend constants, p_0 the undisturbed pitch and d the layer thickness.

6.3. Texture Change Effect (Memory Effect)

The texture change or memory effect is observed in cholesteric materials with negative dielectric anisotropy [71]. The liquid crystal layer is homogeneously oriented by boundary forces to form the planar texture which is completely transparent if the band of selective light reflection is outside the visible spectrum. The substrates are covered with conducting films that are in contact with the liquid crystal. When a d.c. or low frequency field is applied, the sample is transformed to the so-called focal conic texture. In this texture, the liquid crystal is broken up into small domains which are randomly oriented and have diameters of a few microns. Since these domains are optically anisotropic, they act as scattering centers for visible light. Therefore the focal conic texture exhibits a milky white appearance.

The effect is due to the formation of space charges, analogeous to Williams domain formation in nematic liquid crystals. The applied electric field exerts forces on these space charges and thus causes electrohydrodynamic processes that destroy the homogeneous planar texture.

After removing the voltage, the scattering state lasts for hours, days or even weeks, depending on the material used. Therefore the effect can be used in electro-optic memory devices [71]. The transparent state can be restored by applying an a.c. voltage which erases the "stored information". The frequency of this voltage must be higher than the space charge limited dielectric relaxation frequency of the material. The erasure time depends on the applied a.c. voltage. Rather large fields (close to the dielectric breakdown strength) must be used to obtain erasure times smaller than one second.

In another optical memory display scheme the transformation from the planar to the focal conic texture is achieved thermally [72]. An infrared laser beam is used to heat a small spot of the liquid crystal layer above the cholesteric-isotropic transition temperature. When cooling down, this region spontaneously adopts the scattering focal conic texture. The initial transparent state is restored by applying an a.c. field as described above.

References

1. Kast, W.: Angew. Chem. **67**, 592 (1955).
2. Brown, G. H., Shaw, W. G.: Chem. Revs. **57**, 1049 (1957).
3. Chistyakov, I. G.: Soviet Physics Uspekhi **9**, 551 (1967).
4. Saupe, A.: Angew. Chem. (Internat. Edit.) **7**, 97 (1968).
5. Sackmann, H., Demus, D.: Fortschr. Chem. Forsch. **12**, 349 (1969).
6. Baessler, H.: In: *Festkörperprobleme XI – Advances in Solid State Physics* (O. Madelung, Ed.) Oxford and Braunschweig: Pergamon and Vieweg, p. 99, 1971.
7. Brown, G. H.: J. Opt. Soc. Am. **63**, 1505 (1973); Brown, G. H., Doane, I. W.: Appl. Phys. **4**, 1 (1974).
8. Gray, G. W.: *Molecular Structure and the Properties of Liquid Crystals*, London: Academic Press 1962.
9. Gray, G. W., Winsor, P. A.: Eds. *Liquid Crystals and Plastic Crystals,* Chichester: Ellis Horwood Ltd. 1974.
10. de Gennes, P. G.: *The Physics of Liquid Crystals,* Oxford University Press 1974.
11. Friedel, G.: Ann. Physique **18**, 273 (1922).
12. Demus, D., Klapperstück, M., Link, V., Zaschke, H.: Mol. Cryst. Liqu. Cryst. **15**, 161 (1971).
13. Maier, W., Saupe, A.: Z. Naturforsch. **13a**, 564 (1958); **14a**, 882 (1959); **15a**, 287 (1960).
14. de Jeu, W. H., van der Veen, J.: Philips Res. Repts. **27**, 172 (1972).
15. Steinsträßer, R., Pohl, L.: Z. Naturforsch. **26b**, 577 (1971).
16. Van der Veen, J., de Jeu, W. H., Grobben, A. H., Boven, J.: Mol. Cryst. Liq. Cryst. **17**, 291 (1972).
17. Steinsträßer, R., Pohl, L.: Tetrahedron Lett. 1971, 1921.
18. McCaffrey, M. T., Castellano, J. A.: Mol. Cryst. Liq. Cryst. **18**, 209 (1972).
19. Young, W. R., Aviram, A., Cox, R. J.: Angew. Chem. **83**, 399 (1971); Angew. Chem. (Internat. Edit.) **10**, 410 (1971); J. Amer. Chem. Soc. **94**, 3976 (1972).
20. Malthete, J., Leclerq, M., Gabard, J., Billard, J., Jacques, J.: C. R. Acad. Sci. C **273**, 265 (1971).
21. Young, W. R., Haller, I., Aviram, A.: Mol. Cryst. Liq. Cryst. **13**, 357 (1971).

22. Kelker, H., Scheurle, B.: J. Physique **30**, C 4-104 (1969); Angew Chemie **81**, 903 (1969); Angew. Chemie (Internat. Edit.) **9**, 962 (1970).
23. Steinsträßer, R., Pohl, L.: Z. Naturforsch. **26b**, 87 (1971).
24. Steinsträßer, R., Z. Naturforsch. **27b**, 774 (1972).
25. Gray, G. W., Harrison, K. J., Nash, J. A.: Electron. Lett. **9**, 130 (1973).
26. Castellano, J. A.: RCA Review **33**, 296 (1972); Ferroelectrics **3**, 29 (1971).
27. Kahn, F. J., Taylor, G. N., Schonhorn, H.: Proc. IEEE **61**, 823 (1973); F. J. Kahn, Appl. Phys. Lett. **22**, 386 (1973).
28. Haller, I.: Appl. Phys. Lett. **24**, 349 (1974).
29. Berreman, D. W.: Phys. Rev. Lett. **28**, 1683 (1972).
30. Wolff, U., Greubel, W., Krüger, H.: Mol. Cryst. Liq. Cryst. **23**, 187 (1973).
31. Saupe, A., Maier, W.: Z. Naturforsch. **16a**, 816 (1961).
32. Zocher, H.: Z. Physik **28**, 790 (1927).
33. Oseen, C. W.: Arkiv Matematik Astron. Fysik **A19**, 1 (1925), Fortschr. Chem. Physik u. Physik Chem. **20**, 1 (1929), Trans. Faraday Soc. **29**, 883 (1933).
34. Frank, F. C.: Disc. Faraday Soc. **25**, 19 (1958).
35. Maier, W., Meier, G.: Z. Elektrochem. **65**, 556 (1961); Z. Naturforsch. **16a**, 470 (1961); Z. Naturforsch. **16a**, 262 (1961).
36. Schadt, M.: J. Chem. Phys. **56**, 1494 (1972).
37. Boller, A., Scherrer, H., Schadt, M., Wild, P.: Proc. IEEE **60**, 1002 (1972).
38. Ashford, A., Constant, J., Kirton, J., Raynes, E. P.: Electron. Lett. **9**, 118 (1973).
39. Meier, G., Saupe, A.: Mol. Cryst. **1**, 515 (1966); Martin, A., Meier, G., Saupe, A.: Symposium of the Faraday Soc. No. 5, 119 (1971).
40. Berreman, D. W., Scheffer, T. J.: Phys. Rev. **A5**, 1397 (1972); Phys. Rev. Lett. **25**, 577 (1970); Mol. Cryst. Liq. Cryst. **11**, 395 (1970); Berreman, D. W.: J. Opt. Soc. Am. **62**, 502 (1972).
41. Oseen, C. W.: Trans. Faraday Soc. **29**, 833 (1933).
42. De Vries, Hl.: Acta Cryst. **4**, 219 (1951).
43. Dreher, R., Meier, G., Saupe, A.: Mol. Cryst. Liq. Cryst. **13**, 17 (1971).
44. Dreher, R., Meier, G.: Sol. State Comm. **13**, 607 (1973); Phys. Rev. **A8**, 1616 (1973).
45. Kassubek, P., Meier, G.: Mol. Cryst. Liq. Cryst. **8**, 305 (1969).
46. Meyer, R. B.: Phys. Rev. Lett. **22**, 918 (1969).
47. Helfrich, W.: Z. Naturforsch. **26a**, 833 (1971), Phys. Lett. **35A**, 393 (1971).
48. Schmidt, D., Schadt, M., Helfrich, W.: Z. Naturforsch. **27a**, 277 (1972).
49. Saupe, A.: Z. Naturforsch. **15a**, 815 (1960).
50. Gruler, H., Scheffer, T. J., Meier, G.: Z. Naturforsch. **27a**, 966 (1972).
51. Deuling, H.: Mol. Cryst. Liq. Cryst. **19**, 123 (1972).
52. Schiekel, M. F., Fahrenschon, K.: Appl. Phys. Lett. **19**, 391 (1971).
53. Kahn, F. J.: Appl. Phys. Lett. **20**, 199 (1972).
54. Soref, R. A., Rafuse, M. J.: J. Appl. Phys. **43**, 2029 (1972).
55. Assouline, G., Hareng, M., Leiba, E.: Electron. Lett. **7**, 699 (1971); Assouline, G., Hareng, M., Leiba, E., Roncillat, M.: Electron. Lett. **8**, 45 (1972); Hareng, M., Leiba, E., Assouline, G.: Mol. Cryst. Liq. Cryst. **17**, 361 (1972); Hareng, M., Assouline, G., Leiba, E.: Proc. IEEE **60**, 913 (1972).
56. Schadt, M., Helfrich, W.: Appl. Phys. Lett. **18**, 127 (1971).
57. Berreman, D. W.: J. Opt. Soc. Am. **63**, 1374 (1973); Appl. Phys. Lett. **25**, 12 (1974).
58. Van Doorn, C. Z., Heldens, S. L. A. M.: Phys. Letters **47A** 135 (1974).
59. Baur, G., Meier, G.: Vth International Liquid Crystal Conference, June 17–21, 1974, Stockholm, Sweden; Phys. Letters **50A**, 149 (1974). Baur, G.: 1974 Conference on Display Devices and Systems, October 9–10, New York, USA, 1974; Baur, G., Windscheid, F., Berreman, D. W.: Appl. Phys. **8**, 101 (1975).
60. Heilmeier, G. H., Zanoni, L. A., Barton, L. A.: Proc. IEEE **56**, 1162 (1968).
61. Björnstahl, Y.: Ann. Physik **56**, 161 (1918).
62. Williams, R.: J. Chem. Phys. **39**, 384 (1963).
63. Carr, E. F.: Mol. Cryst. Liq. Cryst. **7**, 253 (1969).
64. Helfrich, W.: J. Chem. Phys. **51**, 4092 (1969).

65. Leslie, F. M.: Quart. J. Meck. Appl. Math. **19**, 357 (1966); Arch. Rational Mech. Anal. **28**, 265 (1968).
66. Greubel, W., Wolff, U.: Appl. Phys. Lett. **19**, 862 (1972).
67. de Gennes, P. G.: Solid State Comm. **6**, 163 (1968).
68. Meyer, R. B.: Appl. Phys. Lett. **12**, 281 (1968).
69. Jakeman, E., Raynes, E.: Phys. Lett. **A39**, 69 (1972).
70. Leslie, F.: Mol. Cryst. Liq. Cryst. **12**, 57 (1970).
71. Heilmeier, G., Goldmacher, J.: Appl. Phys. Lett. **13**, 132 (1968); Proc. IEEE **57**, 34 (1969).
72. Melchior, H., Kahn, F., Maydan, D., Fraser, D.: Appl. Phys. Lett. **21**, 392 (1972).

Scientific Applications of Liquid Crystals

E. Sackmann
Abteilung für experimentelle Physik III der Universität Ulm,
D-7900 Ulm, Federal Republic of Germany

Introduction

Liquid crystals are excellent solvents for organic molecules. Non-mesomorphic guest molecules may be incorporated in liquid crystals up to fairly high concentrations without destruction of the distant order prevailing in the liquid crystalline matrix. The guest molecules couple to the anisotropic intermolecular interaction field of the oriented host but can diffuse rather freely within the solvent. The anisotropic guest-host interaction leads to a highly anisotropic tumbling motion of nonspherical guest molecules within the liquid crystal matrix. In the time average this anisotropic tumbling leads to an appreciable orientation of the guest with respect to the axis of preferred solvent alignment. The liquid crystalline solutions can be oriented homogeneously by electric, magnetic or mechanical forces, that is bulk samples of highly oriented solute molecules can be easily prepared.

The molecular properties of the guests are not altered appreciably by the weak intermolecular forces. Accordingly, guest molecules incorporated into liquid crystals behave as an "oriented gas". This finding provides the basis for the applicability of liquid crystals as anisotropic solvents for spectroscopic investigations of anisotropic molecular properties. As will be shown in Sections 2 and 3 spectroscopic studies in liquid crystals provide a great deal of additional information on the physical properties of molecules.

As a byproduct of all spectroscopic studies one obtains values of the average orientations of the solute molecular axis with respect to the liquid crystal optic axis. These average solute orientations (the so-called order parameters of the solute) are closely related to the anisotropic potential of the guest molecule in the aniso-tropic molecular field produced by the liquid crystals. Solutions in liquid crystals may therefore be considered as model systems for investigations of the anisotropic intermolecular Van-der-Waals forces. Such applications of the liquid crystals will be described in Section 4.

The anisotropic solute—solvent interaction depends critically on the geometry of the guest molecules and therefore provides a very sensitive physical parameter to distinguish between two geometrical isomers of a molecule. For this reason liquid crystals may be used most successfully as substrates in gas-liquid chromatography. This type of application is described in Section 5.

Small concentrations of guest molecules may induce dramatic changes in the pitch of cholesteric liquid crystals. The corresponding (reflection) color response may be exploited in analytical chemistry as described in Section 6.

1. Solvents and Methods of Orientation

1.1. Solvents

Most known mesomorphic compounds are aromatic systems of the following general shape

$$\text{(I)}$$

The bridge is provided by an unsaturated functional group such as $-N = N(O) -$; $-N = CH -$; $-N = N -$; $-CH = CH -$. Some of the most widely used liquid crystals of this type are summarized in Table 1.

Since the pioneering synthesis of MBBA (Ia) by Kelker and Scheurle [1]

$$\text{(Ia)}$$

several liquid crystals of this type which are nematic at room temperature have become available. Most of these room temperature nematics are, however, eutectic mixtures of two or more low melting compounds [2, 3, 4]. The development of low melting nematic phases has considerably stimulated the application of liquid crystals as anisotropic solvents.

Most of the above mesomorphic compounds are not suited as solvents for polar molecules such as amino acids. The most polar liquid crystalline solvents of type (I) are provided by derivatives of benzoic acids, such as 4-n-octyloxy benzoic acid

[5] (Ib). These compounds tend to form elongated (rod-like) dimers in which the molecules are held together by hydrogen bonds formed between the carboxyl groups of two molecules.

(Ib)

Many of the compounds of type I exhibit one or several smectic phases besides a nematic phase. It has been shown recently [6], that smectic liquid crystals can in principle be aligned homogeneously by cooling an ordered nematic phase below the nematic-smectic transition temperature. Accordingly, smectic phases are also suited as anisotropic solvents for spectroscopic applications.

The applicability of the liquid crystals of type I for optical experiments is limited due to the strong absorption of these molecules in the near ultraviolet (uv) wavelength region. The cholesteryl derivatives (II)

(II)

are transparent to about 230 nm [7]. The pure cholesteryl derivatives are cholesteric or smectic and cannot be aligned homogeneously over large volumes. Bulk samples of these compounds can be ordered by using so called "compensated" nematic mixtures that are obtained by mixing cholesteryl chloride (or cholesteryl bromide) with cholesteryl esters in appropriate amounts. The Pitch P of these mixtures is strongly temperature-dependent and there exists a characteristic temperature, T_{nem}, where the mixture is nematic (that is where the pitch P is infinity). In the neighborhood of T_{nem} the pitch P is inversely proportional to the absolute value of the temperature difference $|T - T_{nem}|$ as shown in Fig. 1.

By using suitable cholesteryl derivatives and by varying the composition, nematic mixtures can be prepared between $10°C$ and $150°C$. Several "compensated" mixtures of cholesteryl chloride with some cholesteryl esters and with one achiralic compound, namely 4,4'-di-n-hexyloxyazoxybenzene (Ic), are summarized in

(Ic)

Fig. 2 where T_{nem} is plotted as a function of the molar percentage of cholesteryl chloride. It is seen (1) that T_{nem} decreases linearly with increasing cholesteryl chloride content and (2) that the nematic state of the mixture can be shifted over a large temperature region by only minor changes in the composition.

The true melting points of the mixtures shown in Fig. 2 are well above room

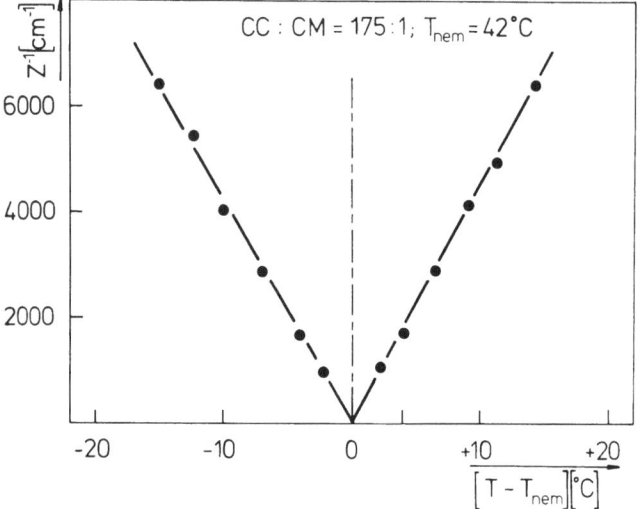

Figure 1. Temperature dependence of the pitch of a 1.75:1 by weight mixture of cholesteryl chloride ($X \equiv C\ell$, CC) and cholesteryl myristate ($X \equiv CH_3 (CH_2)_{12} COO-$, CM) T_{nem} = 42°C. Z is the distance between two planes of equal molecular alignments (Z = P/2). (Values of Z taken from Ref. 8).

temperature. Fortunately, most of these nematic mixtures undercool readily and remain stable in the (metastable) nematic state for a fairly long period of time. A 1.85 : 1 by weight mixture of cholesteryl chloride and cholesteryl laurate, for instance, is nematic at 30°C for several hours before it crystallizes slowly.

The compensated mixtures provide nematic phases with a local screw sense which orient d- and l-isomers of optically active molecules somewhat differently.

Poly -γ- benzyl -L- glutamate: α- helix structure

$R \equiv -(CH_2)_2 - COOCH_2 - \bigcirc$

(III)

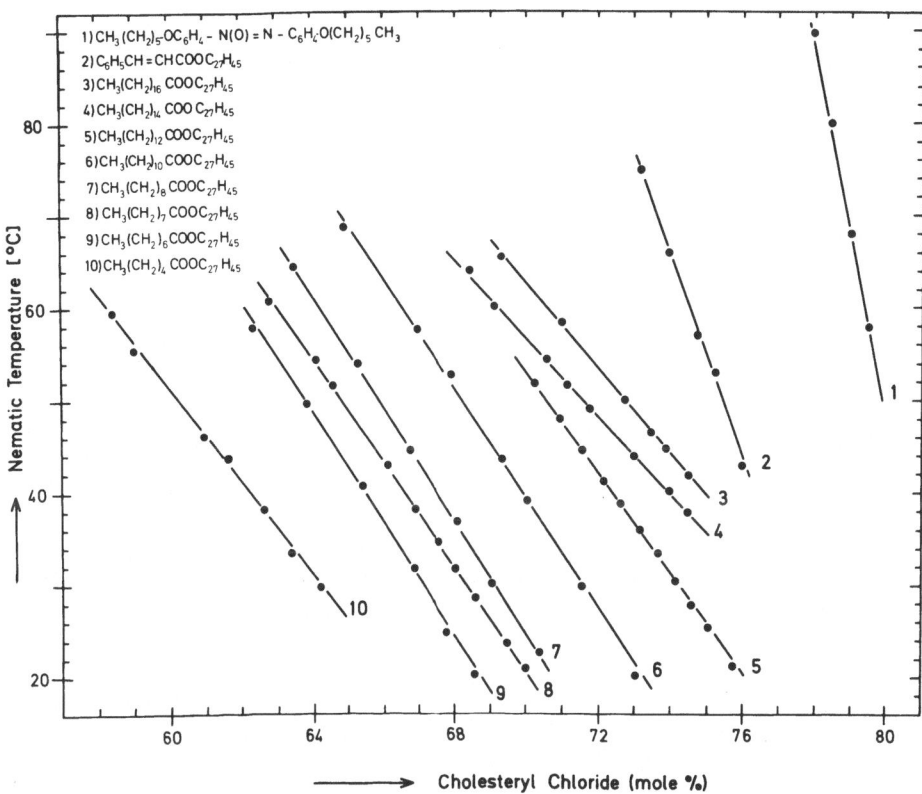

Figure 2. Concentration dependence of the nematic temperature "T_{nem}" for mixtures of cholesteryl chloride with several cholesteryl esters and with 4,4'-hexyloxyazoxybenzene (Ic). The nematic temperature T_{nem}, is plotted as a function of the molar percentage of cholesteryl chloride.

This possibility may be of great value for the spectroscopic investigation of chiralic molecules. An example will be given in Section 2.

Partially compensated cholesteric and compensated nematic solvents are also obtained by dissolving rod-like polymers, such as poly-γ-benzyl-L-glutamate (III) in the α-helical conformation, in organic solvents (methylene chloride, dioxane). The mesomorphic behavior of these solutions (or lyotropic liquid crystal) was first described in detail by Robinson [9]. Applications in nmr spectroscopy have been described by Panar and Phillips [10] and by Samulski and Tobolsky [11].

A most interesting lyotropic nematic solvent which is suitable as anisotropic solvent for very polar molecules has been proposed by Lawson and Flautt [12], and Black et al. [13]. A typical nematic phase of this type is composed of 35 wt% sodium decyl sulfate, 7 wt% n-decyl alcohol, 7 wt% sodium sulfate and 51 wt% water (D_2O in nmr experiments). This solvent is stable between 10°C and 75°C. Moreover, these mixtures can be easily oriented homogeneously by magnetic fields.

For some spectroscopic applications lyotropic lamellar mesophases of soaps may be of great value. Lamellar liquid crystals are formed in sodium decanoate-decanol-water systems [14]. A convenient system is obtained by mixing 28 wt% sodium

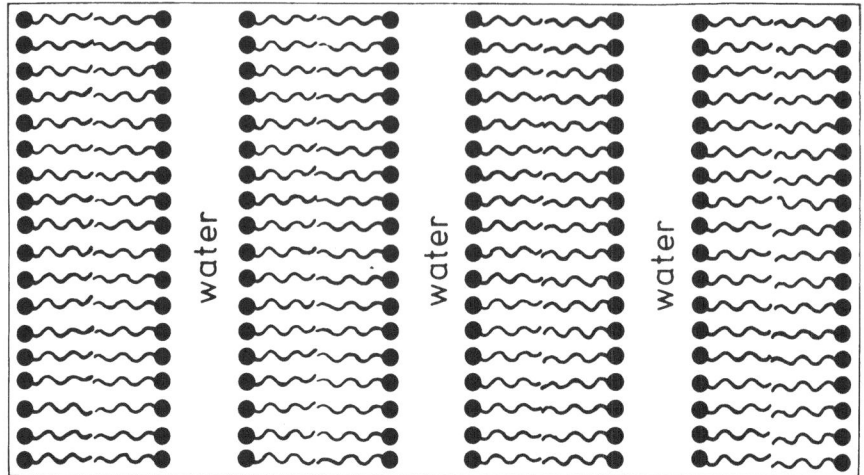

Figure 3. Lamellar phase of soap−alcohol−water dispersions. Projection in a direction parallel to the membrane surfaces. The closed circles represent the polar groups. A typical lamellar phase contains 28 wt% sodiumdecanoate ($CH_3(CH_2)_8COONa$), 42 wt% decanol ($CH_3(CH_2)_9OH$) and 30 wt% water.

decanoate, 42 wt% decanol and 30 wt% water [15]. In the lamellar phase (Fig. 3) bimolecular layers of about 25 Å thickness are formed, where the polar head groups form an interface with water. The hydrocarbon chains directed towards each other form a highly anisotropic hydrophobic phase. Since the bilayers are separated by water they are easily movable against each other. Accordingly the lamellar systems may be oriented uniformly by hydromechanical shear between parallel glass plates separated by about 0.05 cm [15].

In the neighborhood of the semipolar interface the average orientation of the alkyl chains is characterized by an order parameter of S ~ 0.65. Accordingly these lamellar phases are well suited to prepare highly ordered solutions of polar and amphiphilic molecules.

1.1.1. Solute Induced Narrowing of Mesomorphic Range

An intrinsic problem in the application of liquid crystals as anisotropic solvents is the strong suppression of the nematic-isotropic transition temperature (T_i) by the solute molecules. Since, in general, the melting point is much less suppressed by the solute than the value of T_i the usable mesomorphic range may be prohibitively narrowed if high solute concentrations are necessary. Moreover the solvent order may be diminished at high solute concentration. The isotropic nematic transition decreases linearly with increasing solute concentration [16]. Addition of 0.2 M ethanol and hexadecanol, respectively, to MBBA depress the transition temperature T_i by 10°C and 20°C, respectively. 0.9 M allene in n-hexyloxyazoxybenzene decreases the transition temperature T_i by 40°C and the melting point by 10°C. These examples show that rather high concentrations of small molecules can be incorporated in nematic liquids. This finding is most important for applications of liquid crystals in high resolution nmr spectroscopy. For most spectroscopic studies, however, much lower solute concentrations are sufficient and the narrowing of the mesomorphic range is not critical in most cases.

1.1.2. Purification of Liquid Crystals

A severe problem is presented by the purification of liquid crystalline solvents. Especially for applications in optical spectroscopy extremely (spectroscopically) pure solvents are often desirable. To the author's knowledge spectroscopically pure liquid crystals have not yet been prepared. The cholesteryl derivatives contain impurities which absorb between 330 nm and 270 nm and which emit a broad structureless emission band centered at 400 nm. Most probably these impurities are steroids with more than one double bond. In the author's laboratory the purest cholesteryl derivatives have been obtained by recrystallization first from etanol and subsequently from dioxane. By this procedure the impurity absorption could be reduced by a factor of 10.

The conductivity of liquid crystals is rather high (typically of the order of 10^{-12} Ω^{-1} cm^{-1}), due to the presence of ionic impurities. Accordingly the orientation of liquid crystals by electric fields is often disturbed by current-induced fluctuations of the local optic axes in the liquid crystalline matrix. The current-induced fluctuations can be considerably suppressed by removing the ionic impurities by the method of electrodialysis [17, 18]. The electrodialysis of liquid crystals is easily effected in electrolytic cells by covering the electrodes of the cell with semipermeable membranes. For this purpose the anode is covered by a membrane (type A 60, American Machine and Foundry Company) which is selectively permeable to negatively charged impurities while the kathode is covered with a membrane (type C-60) permeable to positively charged impurities only. The conductivity can be decreased by about three orders of magnitude after 2 hours of electrodialysis at 500 V/cm.

1.2. Methods of Solvent Alignment

During the last years a considerable number of techniques for the preparation of homogeneously oriented liquid crystals has been developed. The spectroscopist who wants to use an anisotropic solvent should therefore in most cases find both a suitable liquid crystal and a convenient technique of orientation. In the following the most important techniques of solvent alignment are summarized.

1.2.1. Ordered Layers of Liquid Crystals

Ordered layers of nematic phases up to about 0.1 mm thickness can be prepared by sandwiching the liquid crystal between flat glass plates. For applications in infrared spectroscopy NaCl-plates may also be used as substrate. Depending on the pretreatment of the substrate the optic axis of the nematic phase may be oriented either parallel (homogeneous orientation) or perpendicular (homeotropic orientation) to the surface of the substrate.

Homogeneous Alignment

Homogeneous alignment of nematic phases is achieved by rubbing the surface of the substrates with paper, cloth or leather. The molecules (or the directors) of the liquid crystals tend to align parallel to the direction in which the surfaces have been rubbed. Excellent alignment is also achieved by grinding the glass surfaces with diamond paste [19]. Homogeneous oriented nematic layers have been used for uv and infrared spectroscopic studies [20, 21].

Cholesteric phases can also be ordered between glass plates. By the application of a small pressure a so called layered or "Grandjean" texture is obtained: in this texture the nematic layers of the cholesteric phase orient parallel to the surface, that is the helix (or optic) axis points in a direction normal to the glass surface. Such oriented cholesteric layers have been used for circular dichroism studies (Section 3).

Homeotropic Alignment

Homeotropic alignment can be achieved for instance by cleaning the glass surfaces carefully with chromosulphuric acid [22]. Good results are obtained if the sample is first heated above the nematic-to-isotropic transition and is then cooled slowly to the nematic state. Nematic phases of type I (such as MBBA) are well suited for homeotropic orientation. A homeotropic orientation is most easily achieved if the substates are covered by a thin layer of egg-lecithin [23]. For this purpose the two substrates are dipped into an etheric solution of the lecithin. Using glass plates covered with lecithin the "compensated" nematic mixtures of the cholesteryl[1] derivatives may also be oriented homeotropically. In fact, the tendency of the molecules to orient perpendicular to the glass surface is so strong that a slightly twisted cholesteric phase is transformed into an untwisted nematic texture [24].

Smectic liquid crystals may also be oriented homeotropically. By sandwiching smectic phases between flat glass plates (separated by spacers of about $10-50 \, \mu m$ thickness) the layers are oriented parallel to the surface, i.e. the molecules are oriented normal to the surface.

As mentioned above, the lyotropic lamellar phases of soap-alcohol-water mixtures [15] or of phospholipid-water dispersions may also be ordered homeotropically. Ordered layers up to 0.5 mm thickness may be produced. The orientation is effected simply by sucking the dispersions into flat glass cells with an inner diameter $0.5 \times 10.0 \, \text{mm}^2$. The torque generated by the hydrodynamic shear orients the membranes in such a way that their normals are directed perpendicular to the glass surface.

1.2.2. Magnetic Field Orientation

Nematic Phases

Nematic liquid crystals can be easily oriented by static magnetic fields. Two cases have to be distinguished depending on the sign of the diamagnetic anisotropy $(\chi_\parallel - \chi_\perp)$ of the molecules forming the liquid crystal.

$(\chi_\parallel - \chi_\perp) > 0$: (*Positive diamagnetic anisotropy*). The aromatic nematogenes of type I are examples of compounds with positive diamagnetic anisotropy. The long axes of the molecules tend to align parallel to the magnetic field. The nematic liquid crystals orient in the magnetic field $\vec{H}$ in such a way that the directors of the spontaneously ordered domains are all aligned uniformly in one direction (parallel to $\vec{H}$). In general, field strengths of several hundred Gauss are sufficient to produce uniform orientation. The magnetic field strengths used in nuclear magnetic (nmr) and electron spin (esr) resonance spectroscopy are therefore more than sufficient to produce a homogeneous orientation. Distortions may occur near the surface of the glass walls if the direction of the torque generated by the glass wall does not coincide with the direction of the torque exerted by the magnetic field. Fortunately the transition region of distortion is very thin. Its thickness is inversely proportional to the magnetic field strength H [28] and is of the order of 2–5 μm at $H = 10^4$ Gauss.

   $(\chi_\parallel - \chi_\perp) < 0$: (*Negative diamagnetic anisotropy*). Typical examples of nematogenes of this type are the cholesteryl derivatives [26], the nematic solutions of alkyl sulphates introduced by Lawson and Flautt [12] and the concentrated solutions of poly-γ-benzyl-L-glutamate [9].

   The molecules tend to align perpendicular to the magnetic field direction. The directors of the ordered domains become oriented perpendicular to the field $\vec{H}$. They may, however, point in any direction perpendicular to $\vec{H}$. Accordingly, the nematogens with negative diamagnetic anisotropy do not orient uniformly in a magnetic field, but exhibit a lower order with respect to an external axis than the nematic phases with positive anisotropy. Nevertheless, it is possible to obtain high resolution nmr spectra in liquid crystalline solvents of this type [26]. High resolution nmr spectra have also been observed in the nematic soap solution [13].

Cholesteric Phases

Cholesteric liquid crystals composed of molecules with negative diamagnetic anisotropy ($|\chi_\parallel| > |\chi_\perp|$) may also be ordered uniformly by magnetic fields. Since molecules of this type tend to orient perpendicular to the magnetic field direction, the helix axes of the spontaneously ordered cholesteric domains are turned in a direction perpendicular to the magnetic field. Such an experiment will be described in more detail below.

Smectic Phases

Very recently Meiboom and Luz [6] have shown that smectic liquid crystals of compounds which exhibit a nematic and a smectic phase may be oriented uniformly by a magnetic field in the following way; first the nematic phase is oriented in the magnetic field. Then the sample is slowly cooled below the nematic-to-smectic transition while it is kept in the magnetic field. In this way all three smectic phases A, B, C of terephtal-bis-(4-n-butylaniline) can be ordered uniformly.

1.2.3. Electric Field Orientation

In an electric field $\vec{E}$ a torque is generated which tries to orient the spontaneously ordered domains in such a way that the axis of maximum dielectric polarizability is parallel to the field direction. Again two cases have to be distinguished.

Negative Dielectric Anisotropy ($\epsilon_\parallel - \epsilon_\perp < 0$)

Most of the aromatic nematogenes of type I belong to this class (cf Table 1). The directors, $\vec{L}$, of the nematic domains tend to align perpendicular to the electric field $\vec{E}$. At weak d.c. electric fields bulk samples can be aligned at least partially. At high field strengths, that is above the so called threshold voltage (Part I and III of this volume), the order is destroyed again by the onset of fluctuations of the local directors (range of dynamic scattering). These fluctuations are current-induced. Accordingly, much better alignment can be achieved by using high frequency alternating electric fields. Frequencies larger than 500 Hz are sufficient to suppress the fluctuations considerably. The fluctuations may also be suppressed if the liquid crystal is purified carefully by electrodialysis.

Positive Dielectric Anisotropy ($\epsilon_\parallel - \epsilon_\perp > 0$)

The directors of the spontaneously ordered domains are rotated in a direction parallel to the electric field $\vec{E}$. A uniformly oriented state is obtained which is stable at all voltages. Accordingly, homogeneously ordered bulk samples can be easily prepared in d.c. fields. Examples for nematogens with positive anisotropy are the compensated mixtures of the cholesteryl derivatives. Such ordered phases have been used extensively for optical polarization experiments in the ultraviolet wavelength region (cf Ref. 7). Two examples for type I nematogenes with positive dielectric anisotropy are given in Table 1. Field strength of the order of 10,000 V/cm are sufficient to produce good alignment. Due to ionic impurities present in unpurified liquid crystals, the field strength of 10,000 V/cm induces rather high current densities (of the order of 10 μA/cm^2). First these currents involve inhomogeneities in the electric field $\vec{E}$ both in a direction parallel and perpendicular to $\vec{E}$. Secondly, the currents induce considerable fluctuations in the director of the liquid crystal. Both effects may be suppressed (1) by purifying the liquid crystal using electrodialysis and (2) by using high frequency alternating fields.

1.2.4. Cholesteric-to-nematic Transition

Molecules with positive diamagnetic anisotropy ($|x_\parallel| < |x_\perp|$) or with positive dielectric anisotropy ($\epsilon_\parallel > \epsilon_\perp$) tend to align parallel to the direction of a magnetic or an electric field, respectively. Cholesteric phases that are composed of such molecules should therefore transform into nematic phases under strong magnetic or electric fields Consider the case of an electric field; it is clear that unwinding can occur if the gain in electric energy ($\epsilon_\parallel - \epsilon_{chol})E^2/8\pi$, overcompensates the elastic energy of the helical twisting. The latter is given by

$$E_{el} = \frac{1}{2} k_{22} \left(\frac{2\pi}{P}\right)^2, \tag{1}$$

where k_{22} is the twist elastic constant [28]. Obviously, the helical structure becomes unstable above a critical field strength E_c which is defined as that field where the consumption in elastic twist energy and the gain in electric energy compensate:

$$E_c = \frac{2\pi}{P} \left(\frac{4\pi k_{22}}{\epsilon_\parallel - \epsilon_{chol}}\right)^{\frac{1}{2}} \tag{2}$$

If the helix axis $\vec{h}$ was originally oriented perpendicular to the field direction it is $\epsilon_{chol} = \frac{1}{2} (\epsilon_\parallel + \epsilon_\perp)$.

Field-induced cholesteric → nematic transitions were first observed by Sackmann et al [26]. in a magnetic field and by Wysocki et al. [29] in an electric field. An exact theory of the field-induced unwinding of a helical structure was first given by Meyer [30] and by de Gennes [31]. The first quantitative check of these theories by Bässler and Labes [32] showed good agreement between theoretical and experimental results.

The field-induced cholesteric-nematic transition is most important for the application of "compensated" mixtures of cholesteryl derivatives as anisotropic solvents. If electric fields of the order of 10,000 V/cm are used for the orientation of the liquid crystal it is not necessary to adjust the temperature, T, of the sample exactly to the characteristic nematic temperature, T_{nem}, of the mixture. Good alignment can still be achieved if $|T - T_{nem}| \leqslant 5°C$.

An electric field induced cholesteric-to-nematic transition is demonstrated in Fig. 4 for a partially compensated cholesteric mixture of cholesteryl chloride and cholesteryl nonanoate. The sample has been sandwiched between two glass plates in such a way that the helix axis is parallel to the glass surface. The distance between two adjacent dark (or bright) lines is therefore a measure for the pitch fo the cholesteric phase. It is clearly seen that the pitch increases with increasing field strength. At a critical field strength of $E_c = 10,000$ V/cm the sample has become nematic with the director oriented parallel to the electric field.

1.2.5. Ordered Glasses of Liquid Crystals

For many spectroscopic applications (especially in magnetic resonance) it is desirable to rotate the optic axis $\vec{L}$ of the ordered liquid crystal with respect to an external axis (such as the external magnetic field). The following methods are conceivable in order to perform such experiments:

Crossed electric and magnetic fields

This method can be applied if one chooses a nematic phase with a positive dielectric anisotropy ($\epsilon_\parallel > \epsilon_\perp$). An electric field of the order of 10,000 V/cm generates a much larger torque on the nematic domains than the magnetic fields (of about 10,000 Gauss) usually used in magnetic resonance experiments. Accordingly, the order generated by an electric field of such a field strength is not destroyed if the sample is rotated in the magnet of a conventional nmr- or esr-spectrometer. This method was first applied by Diehl et al. [33] using 4-methoxy-benzylidene-4-amino-α-methyl cinnamic acid-n-propyl ester (Ic).

(Id)

Use of smectic liquid crystals

As noted above, smectic liquid crystals may be uniformly ordered by cooling a magnetic field-oriented nematic phase below the nematic-to-smectic transition temperature [6].

Due to the very high viscosity of the smectic state the alignment of the ordered smectic phase is not disrupted by thermal fluctuations if the sample is removed from the magnetic field. Moreover the order of the smectic phase is also maintained if the optic axis is rotated with respect to the magnetic field. The same is valid for the lyotropic smectic phases of alcohol-water-soap dispersions ordered by mechanical shear [15].

Rigid ordered glasses

Many mesomorphic compounds undercool readily and may therefore form rigid

Figure 4. Electric field-induced cholesteric-to-nematic transition of a 1.8:1 by weight mixture of cholesteryl chloride and cholesteryl nonanoate. Viewed through crossed polarizers (magnification × 200). The helix axis was parallel and the electric field was perpendicular to the glass plates (separated by 20 μm). The mixture was nematic 5°C below the temperature of observation which was T = 40°C.
a) o V/cm b) 3000 V/cm c) 6000 V/cm d) 10,000 V/cm.

glasses if cooled far below their melting point. The macroscopic order of a homogeneously ordered liquid crystal is often maintained upon cooling. The long range
order of such aligned glasses is extremely stable and cannot be disrupted by an
external field. Such ordered glasses may be of considerable interest for spectroscopic applications. Solute molecules may be incorporated into such glasses up to
fairly high concentrations (0.5 M). In the past, ordered glasses have been used
extensively for phosphorescence polarization experiments (Section 3) and for
esr-spectroscopic studies of the excited triplet states of aromatic molecules as
described in Section 2.

The nematic mixtures of cholesteryl derivatives are excellently suited for the
preparation of ordered glasses. Clear, nearly crack-free glasses may be produced if
the samples are cooled to temperatures between $-20°$C and $-70°$C. At lower
temperature, care must be taken to minimize the number of cracks.

The conservation of the alignment in ordered glasses is demonstrated in Fig. 5.
This figure shows the diffraction pattern of a magnetic field oriented (partially
compensated) cholesteric mixture of cholesteryl chloride and cholesteryl myristate
(1.75 : 1 by weight) taken with a He-Ne laser. The cholesterol derivatives possess
a positive diamagnetic anisotropy $(|\chi_{\parallel}| > |\chi_{\perp}|)$, that is the nematic layers of the
cholesteric phase orient perpendicular to the magnetic field. The magnetic field
oriented cholesteric phase exhibits therefore a periodically varying refractive index
for light that is polarized perpendicular to the helix axis and that is traversing the
liquid crystal in a direction perpendicular to this axis. As expected, the diffraction
pattern of the cholesteric phase is identical to the characteristic diffraction pattern
of an optical grating. The rather clear diffraction pattern of Fig. 5 demonstrates
that the glass is homogeneously ordered on a macroscopic scale.

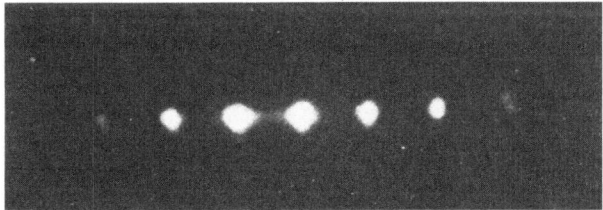

Figure 5. Diffraction pattern of an ordered cholesteric glass (pitch ~ 10 μm) obtained with a
He-Ne-laser (632.8 nm). The light beam was directed perpendicular to the helix axis. The glass
has been prepared by rapid transfer of the sample (cqntained in flat cells of 0.2 mm thickness)
from the orienting magnetic field (30,000 Gauss) to a coolant of $-50°$C (reprinted from
Sackmann et al. [8]).

Homogeneously ordered glasses may also be prepared by cooling electric field
oriented nematic phases of the cholesteryl derivatives. These glasses are optically
uniaxial with the molecules oriented with their long axes parallel to the optic axis.
Electric field-oriented glasses may be prepared as follows: first the liquid crystal
(kept for instance in a glass tube) is oriented in a condenser immersed in a (paraffin)
oil bath that is kept at the nematic temperature of the mixture. By degassing the

Table 1. Some liquid crystal solvents of Type I used for scientific applications.

Compound	Structure	Mesomor [in °C]
N-(p-Methoxy-benzilidene-p-butylaniline (MBBA)	$CH_3O-C_6H_4-CH=N-C_6H_4-C_4H_9$	nem: 20-
p-Azoxyanisole	$CH_3O-C_6H_4-N(O)=N-C_6H_4-OCH_3$	nem: 117
Licrystal V[a]	see below[a]	nem: −5
Butyl-p-(p-Ethoxyphenoxy carbonyl) phenyl carbonate	$C_4H_9O-COO-C_6H_4-COO-C_6H_4-OC_2H_5$	nem: 56-
4,4′-di-Hexyloxyazoxybenzene	$C_6H_{13}O-C_6H_4-N(O)=N-C_6H_4-OC_6H_{13}$	nem: 81- sm: 72−8
4,4′-di-Heptyloxyazoxybenzene	$C_7H_{15}O-C_6H_4-N(O)=N-C_6H_4-OC_7H_{15}$	nem: 92- sm: 74−9
p,p′-Azoxyphenetole	$C_2H_5O-C_6H_4-N=N(O)-C_6H_4-OC_2H_5$	nem: 138 nem: 236
Terephthal-bis-(4-n-butyl-aniline (TBBA)	$C_4H_9-C_6H_4-N=(H)C-C_6H_4-C(H)=N-$ $-C_6H_4-C_4H_9$	sm.A: 19 sm.B: 17 sm.C: 14
4-Methoxy benzylidene-4-amino-α-methyl cinnamic acid-n-propylester	$CH_3O-C_6H_4-CH=N-C_6H_4-CH=C(CH_3)-COOC_3H_7$	

[a]Trade name (Merck, Darmstadt, Germany): composed of a 1:1:1:1 mixture of $CH_3O-C_6H_4-N(O)=N-C$ $C_2H_5O-C_6H_4-N(O)=N-C_6H_4-C_4H_9$ and $C_2H_5O-C_6H_4-N = (O)N-C_6H_4-C_4H_9$

paraffin oil electric fields of 30,000 V/cm may be easily applied. As soon as the orientation has been effected (usually after 2–5 sec) the sample is rapidly transferred to a bath of liquid nitrogen or dry ice. Owing to the high viscosity of the cholesteryl mixtures the order of the liquid crystal is not disrupted during the cooling time.

Ordered glasses may also be prepared by cooling oriented nematic phases of the aromatic nematogenes of type I. If the orientation is effected with a magnetic field the sample may be cooled very slowly while it is in the orienting magnetic field. Provided the viscosity increases smoothly with decreasing temperature, the rigidity of these ordered systems can be varied gradually over many orders of magnitudes.

2. Magnetic Resonance Spectroscopy

2.1. High Resolution Magnetic Resonance Spectroscopy

2.1.1. Introduction

The high resolution nmr spectra of organic molecules in isotropic liquids are determined (1) by the isotropic chemical shift and (2) by the isotropic indirect spin–spin coupling acting between different nuclei of a molecule. The third basic magnetic interaction, the direct dipole–dipole coupling between the nuclear magnetic moments is completely averaged out. Moreover, the solute tumbling motion averages out the anisotropic contributions to the indirect coupling and the chemical shift. The anisotropic magnetic interactions affect the nmr spectra in isotropic media only through their influence of the relaxation times (or the line widths).

The isotropic molecular tumbling in ordinary liquids leads to a loss of most valuable information. Nmr-spectroscopists have therefore been searching for anisotropic solvents which allow the performance of high resolution nmr. Several methods have been proposed: application of electric fields [34], embedding of molecules of interest in stretched polymer sheets or in zeolites [35] and rapid rotation of crystals [36]. The applicability of these methods is strictly limited to special cases. The anisotropy of the basic magnetic interactions has, therefore, been widely ignored until the introduction of nematic liquid crystals as anisotropic solvent by Saupe and Englert in 1963 [37], Englert and Saupe [38] and Saupe [39]. The nmr spectra obtained in liquid crystals are primarily determined by the direct intramolecular spin–spin coupling, whereas the isotropic and anisotropic contribution to the chemical shift and the indirect nuclear spin coupling, respectively, appear as second order effects.

The direct dipole–dipole interaction is a function (1) of the time average over the anisotropic solute tumbling motion and (2) of the internuclear distances. The nmr spectroscopy in liquid crystals may therefore provide information on both the average solute orientation and the molecular geometry. Moreover, information on the absolute signs of the indirect coupling constants, and on the anisotropic contributions to the chemical shift and the indirect coupling, respectively, may be obtained from the nmr-spectra of partially oriented molecules.

A number of excellent reviews on the nmr-spectroscopy in liquid crystals is available now [40, 41, 43, 44]. For a comprehensive review of the literature up to 1969 see the review of Diehl and Khetrapal [43].

2.1.2. Basic Principles of nmr-Spectroscopy of Oriented Molecules

The Spin Hamiltonian

The observed nmr-spectra in uniaxial liquid crystals may be described by the following effective spin Hamiltonian [40, 45]

$$\mathcal{H} = \sum_i \gamma_i (1 - \sigma_i - \sigma_i^{an}) H_{z'} I_{z',i} + \sum_{i<j} \sum J_{ij} \vec{I_i} \vec{I_j} + \sum_{i<j} \sum D_{ij} \left(-\frac{1}{2} I_{x',i} I_{x',j} \right.$$

$$\left. - \frac{1}{2} I_{y',i} I_{y',j} + I_{z',i} I_{z',j} \right) \tag{3}$$

The first term of Equ. (3) accounts for the nuclear Zeeman energy. The second and third term involve both the direct and the indirect spin–spin coupling between all nuclei of the molecules considered. γ_i is the magnetogyric ratio of the ith nucleus, $I_{x',i}$, $I_{y',i}$ and $I_{z',i}$ are the components of the angular momentum operator $\vec{I}$. These components are defined in a space-fixed cartesian coordinate system x', y', z'. $H_{z'}$ denotes the external magnetic field and provides a quantization axis, for the nuclear spin. J_{ij} and σ_i represent the isotropic parts of the spin coupling constants and the chemical shift, respectively, while σ_i^{an} and D_{ij} take account for the corresponding anisotropic contributions to these interaction parameters. σ_i, σ_i^{an}, J_{ij} and D_{ij} are time averages (taken over the anisotropic tumbling motion) of a general shielding tensor σ and a general coupling tensor T [45]. The tensors σ and T are defined in a molecule-fixed coordinate system x, y, z [44, 45]. σ_i, σ_i^{an} and D_{ij} may be expressed in terms of the tensor components of σ and T and of the elements of the Saupe order matrix as shown below.

The anisotropic chemical shift

The isotropic contribution to the chemical shift is given by one third of the trace of the general shielding tensor $(\sigma_i = \frac{1}{3}(\sigma_{xx} + \sigma_{yy} + \sigma_{zz})$. σ_i^{an} may be expressed in terms of the shielding tensor components and the elements of the order matrix. For a molecule with a threefold or higher symmetry axis

$$\sigma_i^{an} = 2/3 S_{zz} (\sigma_{\parallel} - \sigma_{\perp}), \tag{4}$$

where $\parallel$ and $\perp$ refer to a direction parallel and perpendicular to its symmetry axis. The anisotropy of the shielding constant $(\sigma_{\parallel} - \sigma_{\perp})$ may in principle be determined by comparing the chemical shifts (of the molecule of interest) measured in the isotropic and in the anisotropic (nematic) phase of the same solvent. Moreover, S_{zz} has to be determined from the direct coupling constants as obtained from the nmr-spectrum that has been taken in the nematic phase [43].

The Anisotropic Internuclear Interactions

The isotropic part of the internuclear interaction is identical to the well known indirect coupling constant and is given by one third of the trace of the general coupling tensor T. The anisotropic nuclear coupling, D_{ij}, contains both the direct dipole–dipole interaction, D_{ij}^{dir} and the anisotropic part of the indirect nuclear coupling D_{ij}^{ind}, D_{ij}^{ind} has the same orientational dependence with respect to $\vec{H}$ as D_{ij}^{dir} and is therefore called the pseudo-dipolar coupling.

Theoretical estimates indicate however, [46] that D_{ij}^{ind} is very small compared to the direct nuclear coupling at least for H–H- and C–H-interactions. The anisotropic indirect couplings give rise to significant effects in the nmr spectra of fluorated compounds [45]. In C_6F_6, for instance, the values of D_{ij}^{ind} amount to about 10% of the corresponding direct coupling constants [45]. In principle, D_{ij}^{ind} can be estimated from deviations between the measured D_{ij}-values and the values of D_{ij}^{dir} calculated from the known geometry of the molecule [47].

The quantity of primary importance derivable from the observed nmr spectra is the direct coupling constant D_{ij}^{dir} which is directly related to the geometry of the molecule as follows: The direct dipolar coupling between two nuclei i and j of molecules performing a rapid tumbling motion (isotropic or anisotropic) is given by:

$$D_{ij}^{dir} = -K_{ij} \left\langle \frac{3\cos^2\theta_{ij} - 1}{R_{ij}^3} \right\rangle \; ; \; K_{ij} = \frac{\gamma_i\gamma_j h}{4\pi^2} \tag{5}$$

θ_{ij} is the angle between the vector connecting the coupling nuclei and the applied field H_z and R_{ij} is the internuclear distance. The bracket $<\,>$ denotes the time average both over the molecular tumbling motion and over the internuclear motions (vibrations, internal conversions). If R_{ij} is measured in Å and D_{ij}^{dir} in MHz, it follows for protons:

$$K_{ij} = \frac{\gamma_H^2 h}{4\pi^2} = 120\ 067\ \text{Hz Å}^3 \tag{6}$$

For a cylindral molecule (such as acetylene) whose long axis, z, coincides with the internuclear vector $\vec{R}_{ij}$, D_{ij}^{dir} is given by [45]

$$D_{ij}^{dir} = 2K_{ij} \langle R_{ij}^{-3} \rangle S_{zz} \tag{7}$$

provided the magnetic field is parallel to the liquid crystal optic axis $\vec{L}$. S_{zz} is the average orientation of the long molecular axis (of acetylen) with respect to $\vec{L}$.

For complicated molecules one chooses a molecule-fixed coordinate system (x, y, z) in such a way that a minimum number of order parameters S_{ij} are necessary to describe the anisotropic orientation of the molecule in the liquid crystal matrix. D_{ij}^{dir} can be expressed explicitly in terms of the inter-nuclear distances by [43, 45]

$$D_{ij}^{dir} = -2\frac{K_{ij}}{\sqrt{5}} \left\{ + c_{3z^2-r^2} \left[\left\langle \frac{(\Delta z_{ij})^2}{r_{ij}^5} \right\rangle_{Av} - \left(\frac{1}{2}\right) \left\langle \frac{(\Delta x_{ij})^2}{r_{ij}^5} \right\rangle_{Av} - \left(\frac{1}{2}\right) \left\langle \frac{(\Delta y_{ij})^2}{r_{ij}^5} \right\rangle_{Av} \right] \right.$$
$$+ c_{x^2-y^2}\sqrt{3} \left[\left(\frac{1}{2}\right) \left\langle \frac{(\Delta x_{ij})^2}{r_{ij}^5} \right\rangle_{Av} - \left(\frac{1}{2}\right) \left\langle \frac{(\Delta y_{ij})^2}{r_{ij}^5} \right\rangle_{Av} \right] + c_{xz}\sqrt{3} \left[\left\langle \frac{(\Delta x_{ij})(\Delta z_{ij})}{r_{ij}^5} \right\rangle_{Av} \right]$$
$$\left. + c_{yz}\sqrt{3} \left[\left\langle \frac{(\Delta y_{ij})(\Delta z_{ij})}{r_{ij}^5} \right\rangle_{Av} \right] + c_{xy}\sqrt{3} \left[\left\langle \frac{(\Delta x_{ij})(\Delta y_{ij})}{r_{ij}^5} \right\rangle_{Av} \right] \right\}. \tag{8}$$

In this equation $\Delta z_{ij} = z_i - z_j$; $\Delta x_{ij} = x_i - x_j$ and $\Delta y_{ij} = y_i - y_j$, where x_i, y_i, z_i are the coordinates of the ith nucleus in the molecular fixed coordinate system. The angular brackets $<\,>_{Av}$ take account for the fact that the internuclear distances are averages over the internal molecular motions. The above equation provides the basis for the determination of the molecular geometry from an analysis of the nmr spectra of partially oriented molecules.

In nmr spectroscopy the solute orientation is often described by an orientational probability function $P(\theta,\phi)$ [45]. $P(\theta,\phi)$ is a measure for the probability (per unit solid angle) that the applied magnetic field $\vec{H}_0$ (or the optic axis of the liquid crystal) assumes the spherical coordinates (θ, ϕ) in the molecule-fixed coordinate system (x, y, z). $P(\theta, \phi)$ can be expressed in terms of spherical harmonics of second order as follows [45]:

$$P(\theta, \phi) = c_0 + c_{3z^2-r^2} D_{3z^2-r^2} + c_{x^2-y^2} D_{x^2-y^2} + c_{xy}D_{xy} + c_{xz}D_{xz} + c_{yz}D_{yz}$$

The coefficients c_i are called motional constants. They are directly related to the elements of the Saupe order matrix by the following relations, where α is the angle between the liquid crystal optic axis and the applied field direction [43, 45].

$$c_{3z^2-r^2} = \sqrt{5} \cdot S_{zz} \cdot \left(\frac{1}{2}\right)(3\cos^2\alpha - 1),$$

$$c_{x^2-y^2} = \sqrt{\left(\frac{5}{3}\right)} \cdot (S_{xx} - S_{yy})\left(\frac{1}{2}\right)(3\cos^2\alpha - 1),$$

$$c_{xz} = 2\sqrt{\frac{5}{3}} \cdot S_{xz} \cdot \left(\frac{1}{2}\right)(3\cos^2\alpha - 1),$$

$$c_{yz} = 2\sqrt{\frac{5}{3}} \cdot S_{yz} \cdot \left(\frac{1}{2}\right)(3\cos^2\alpha - 1),$$

$$c_{xy} = 2\sqrt{\frac{5}{3}} \cdot S_{xy} \cdot \left(\frac{1}{2}\right)(3\cos^2\alpha - 1),$$

This representation of $P(\theta, \phi)$ is a truncated expansion of a general exponential distribution function described in Section 4. The expansion $P(\vartheta,\phi)$ is therefore only valid for small values of the average solute orientation. This condition is usually fulfilled in nmr spectroscopy where only small molecules can be studied as yet. The number of independent motional constants, c_i, is considerably reduced if the solute molecules exhibit some symmetry.

The orientation of a molecule with a threefold (or greater) axis of symmetry can be described completely by a single motional constant; for a molecule with two perpendicular planes of symmetry (e.g. C_{2v}, D_{2h}-symmetry), two constants c are needed. Three independent constants are necessary in order to describe the orientation of molecules with only one plane of symmetry.

Influence of Internal Molecular Motions

The inter-nuclear distances determined from the direct coupling constants D_{ij} are averages over the vibrational motions of the molecules. In order to compare the structural parameters derived from the nmr spectra with the corresponding values obtained by other techniques, such as electron diffraction or microwave spectroscopy, the effects of the vibrations have to be taken into account. The molecular structure (geometry) derived from the nmr spectra may differ systematically from the structure parameters obtained with other techniques for the following reason. In electron diffraction one measures directly the vibrational averages of the internuclear distances $< R_{ij} >$; the microwave spectroscopy yields the averages $< R_{ij}^{-2} >$, whereas the averages $< R_{ij}^{-3} >$ are measured in nmr spectroscopy. The deviations of the measured internuclear distance R_{ij} from the equilibrium distance, R_e, can be calculated for a diatomic molecule [48]. Denoting the relative deviations, ξ, from the equilibrium distance R_e by $\xi = (R_{ij} - R_e)/R_e$ one obtains with the above methods for the bond length R_{ij}

$$R_{ij}^{\text{el. diffr.}} \approx < R_{ij} > \approx R_e(1 + <\xi>)$$

$$R_{ij}^{\text{microw. Spectr.}} \approx \left(\left\langle\frac{1}{R_{ij}}\right\rangle^2\right) - \frac{1}{2} \approx R_e\left(1 + <\xi> - \frac{3}{2}<\xi^2>\right) \qquad (9)$$

$$R_{ij}^{\text{nmr}} \approx \left(\left\langle\frac{1}{R_{ij}}\right\rangle^3\right) - \frac{1}{3} \approx R_e(1 + <\xi> - 2<\xi^2>)$$

This result suggests that the largest differences are to be expected between the nmr and the electron diffraction results.

The correction for the nuclear vibrations is much more difficult for complex molecules. This problem has not yet been solved satisfactorily. An estimate of the influence of the molecular vibrations is, however, necessary in order to determine the pseudo-dipolar coupling constants. Meiboom and Snyder [44] attempted an estimation of the influence of the molecular vibrations on the dipolar coupling constant D_{ij}^{dir} for benzene. By considering only CH-bond stretching and bending as important, these authors postulated that the vibrations affect primarily the C–H coupling constants rather than the H–H-interactions. Consequently, the C–H bond lengths are about by 1.8% larger than the H–H-distances.

2.1.3. Experimental Procedure and Methods of Spectrum Analysis

Experimental: NMR spectra in nematic solvents can be taken with standard high resolution spectrometers. The magnetic fields used for the nmr experiments suffice to produce a uniform orientation of the liquid crystal. A number of critical factors have to be considered:

A high degree of temperature stability and a negligible temperature gradient in the range of the receiver coils is a necessary condition for high resolution experiments in liquid crystals. The temperature stability is most critical if one wishes to work at temperatures higher than room temperature. The line widths generally are of the order of 5 to 50 Hz and are strongly dependent on the temperature gradient. By using lyotropic nematic liquids [12] or spectrometers with superconducting magnets, sample spinning is possible. The line width may be considerably reduced (to 1–5 Hz) by sample spinning.

The nmr intensities are much lower in nematic solvents than in isotropic liquids, due to the large number of lines caused by partially aligned molecules. Accordingly, one uses generally rather high solute concentrations of the order of 20 mole percent. Much lower concentrations are sufficient if a time averaging equipment is available. Extensive time averaging is a prerequisite for measurements of the direct C–H-coupling constants at natural ^{13}C abundance of 1.1% [40, 44, 52].

In order to avoid disturbing solvent lines a solvent with a large number of protons should be used which yields a uniform background of overlapping lines onto which the sharp lines of the guest are superimposed. P-azoxyanisole, for instance, does produce a discreet background [40]. A number of liquid crystals used for nmr spectroscopy have been summarized in the review by Diehl and Khetrapal [43].

Methods of Spectrum Analysis: For simple molecules the line positions and the intensities can be expressed explicitly in terms of the chemical shifts (isotropic and anisotropic) and the coupling constants D_{ij} and J_{ij}. In these special cases a system of equations may be derived which allows the determination of the coupling parameters (D_{ij} and J_{ij}) from the experimental line positions in a straightforward manner [37, 38, 46]. A number of these simple cases has been summarized by Diehl and Khetrapal [43]. For more complex molecules the spectra can only be analyzed by computer methods, either by a simulation or an iteration procedure.

In the computer simulation procedure one uses the molecular structural parameters and the motional constants as input parameters. Programs are available [40, 45] which compute and plot theoretical spectra for any set of input parameters (bond distances and motional constants). By comparing the calculated with the experimental spectra the molecular geometry can be determined by a trial and error procedure.

2.1.4. Examples of Application

The interpretation of nmr spectra in isotropic media is considerably simplified by the well known rule that the indirect spin—spin-coupling between so called equivalent nuclei (nuclei with identical chemical shift) does not lead to an observable line splitting. This rule is no longer true for the direct nuclear coupling. The proton resonance spectrum of acetylene, for instance, is split up in a doublet by the dipolar coupling between the two equivalent protons. The spectrum of partially aligned benzene exhibits about fifty lines.

The spectrum of n interacting nuclei yields $n(n-1)/2$ interaction parameters D_{ij}. Correspondingly, the number of equations correlating D_{ij} with the molecular geometry can be larger than the number of unknown distances and order parameters. In these cases it is possible to determine the entire molecular structure (relative bond length and bond angles) and all order parameters. The amount of obtainable information about the molecular structure depends on the symmetry of the molecules which determines the number of order parameters needed to describe the molecular orientation. An asymmetric molecule of 4 interacting spins is characterized by 5 motional constants. This leaves only one equation which could provide information on the molecular geometry. An extremely important help and in many cases a prerequisite for a structure determination is the measurement of the C—H-coupling constants from the proton resonance spectrum of ^{13}C-isotopes containing molecules.

The motional constants (or the order parameters) and the bond distances appear always as products in Equ. (8). Accordingly, only ratios of nuclear distances (or relative bond length) can be obtained from the nmr-spectra in anisotropic solvents. Absolute values of the order parameters can only be determined for molecules with known internuclear distances.

Molecules of the form $X-CH_3$ A simple example of a molecular structure determination is the measurement of the H—C—H angles in the methyl groups of molecules $X-CH_3$. The proton resonance spectrum of these molecules is characterized by one direct H—H-coupling constant D_{HH}. The spectrum consists of a triplet with relative intensity ratios 1:2:1. The splitting between two adjacent lines is $3 D_{HH}$. For molecules containing a ^{13}C-isotope (spin quantum number 1/2) in the methyl group each of the triplet lines is split in doublets of separation D_{CH}. The orientation of these molecules is described by only one order parameter (if X is an atom or a cylindrically symmetric group). Accordingly, the ratio

$$D_{HH}/D_{CH} = \frac{\gamma_H R_{CH}^3}{\gamma_C R_{HH}^3}$$

is independent of the order parameter and allows the determination of the H—C—H angle in a straightforward way. Using this method Saupe et al. [49] determined the H—C—H angles for acetonitrile, methanole and methyleneiodide. Comparison of their result with the angle obtained by microwave spectroscopy (Table 2) shows excellent agreement.

Table 2. Comparison of H–C–H angles in methyl groups as determined by liquid crystal nmr and by microwave spectroscopy (after Meiboom and Snyder [40] and Saupe et al. [49])

	Liquid crystal nmr	Microwave
$N{\equiv}C{-}CH_3$	$109°2' \pm 2'$	$109°16'$
CH_3OH	$110°3' \pm 8'$	$109° 2'$
CH_3I	$111°42' \pm 2'$	$111°25'$

Cyclopropane. A simple example for a computer simulation procedure is the analysis of the nmr spectrum of cyclopropane (Fig. 6) by Snyder and Meiboom [50]. The total proton spectrum is shown in Fig. 7a. It consists of two identical subspectra which are symmetrically positioned about the center (denoted by 0 Hz).

Cyclopropane has a plane of symmetry (containing the three carbon atoms) and a threefold symmetry axis. Thus the orientation is described by one motional constant, $c_{3z^2-r^2}$. The distances of the nuclei are completely determined by the distances a, b and c of Fig. 6. By varying the ratios of these distances and the order parameter, agreement between calculated and observed spectra was first achieved for those lines which the positions are independent of the indirect coupling constants J_{ij}. The absolute values of the J_{ij} were obtained from the spectrum in an isotropic solvent. Then the signs of the J_{ij} were varied until complete agreement between observed and calculated spectrum was achieved (cf Fig. 7b).

By recording the spectrum at very high sensitivity Snyder and Meiboom could also observe the spectra of those protons that are attached to ^{13}C-isotopes in natural abundance of 1.1%. The analysis of these spectra yields values of all indirect and direct C–H-coupling constants. The complete set of H–H and C–H-coupling constants of cyclohexane is summarized in Table 3.

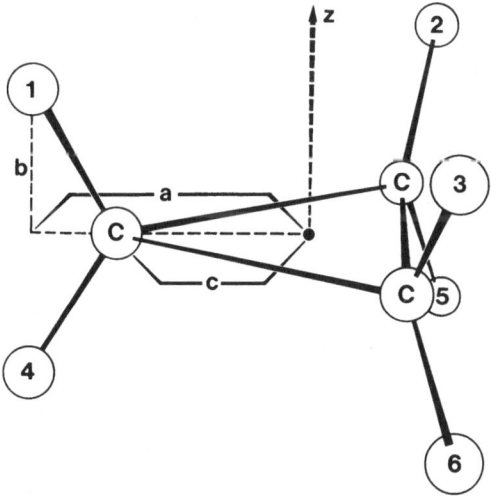

Figure 6. Molecular model of cyclopropane.

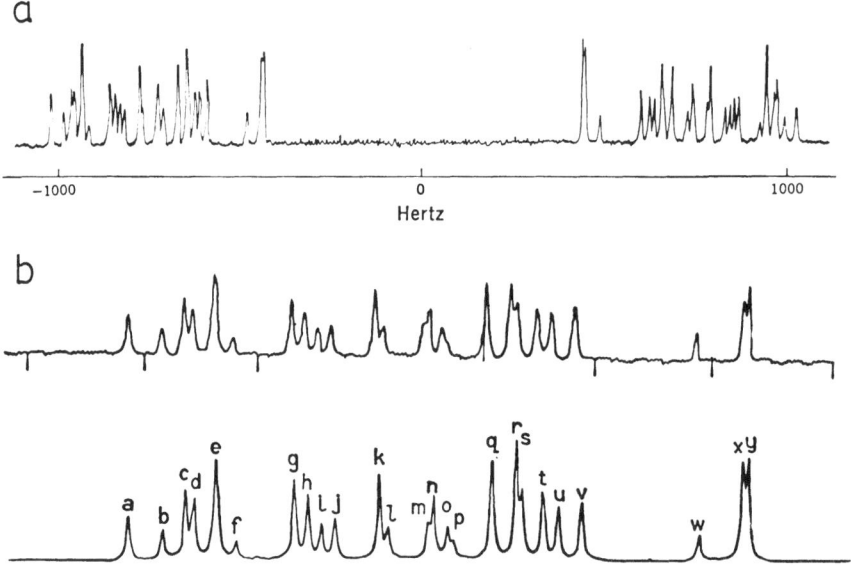

Figure 7. Nmr-spectrum of cyclopropane taken in nematic 4,4-n-hexyloxyazoxybene at 80°C.
(a) Total proton resonance spectrum. Note that the spectrum consists of two identical subspectra that are centered about o Hz.
(b) Comparison of experimental (top) and simulated (bottom trace) spectrum. The spectrum
was simulated with the coupling constants of Table 3 and motional constant $c_{3z}{}^2 - r^2 = -0.03$
(reproduced from Meiboom and Snyder [50]).

The absolute values of the structural parameters given in Table 3 were obtained
by fitting the C–C-bond length to the corresponding value obtained from electron
diffraction [51]. The nmr structural parameters of Table 3 were derived from the
pair of coupling constants $D_{H_1 H_2}$ and $D_{H_1 H_4}$. Somewhat different structural
parameters are obtained if other pairs of coupling constants are used [40].

Compared to the result of electron diffraction the C–C-bond length is too short
relative to the C–H-length by about 0.034 Å. This discrepancy is most probably

Table 3. Coupling constants (in Hertz) and geometry of cyclopropane. The structure parameters have been obtained by fitting D_{12} and D_{14} (values taken from Meiboom and Snyder
[50]).

Pair of nuclei	J_{ij}	D_{ij}	Coordinates	nmr data[a]	Electron[51] diffraction
(1,2)	+9.5	−194.51	C – – –C	(1.510)[a]	1.510
(1,4)	−7.0	+974.00			
(1,5)	+5.5	+4.35	C – – –H	1.123	1.089
(C,1)	+162.0	+651.78	HCH	114.4°	115.1°
(C,4)	0.0	−33.95			

[a]Bond distance fitted to corresponding value obtained by electron diffraction [51].

due to methylene rocking motions. These vibrations shorten the average distance between the protons on the same side of the cyclopropane ring and lead thus to an apparent shortening of the C–C-distance [40].

Allene. As an example of a spectrum that exhibits ^{13}C-satellites Fig. 8 shows one half of the proton nmr spectrum of allene. This spectrum has been recorded after computer accumulation of about 3,000 passes [52]. This example shows that the number of lines is considerably increased when the C–H-coupling is taken into account. The complete analysis of the spectrum of Fig. 8 yields the absolute values and the signs of all direct and indirect H–H- and C–H-coupling constants in a straightforward way. The analysis of the spectrum of protons bound to ^{12}C shows that the planes spanned by the two CH$_2$ groups are oriented perpendicularly. Thus allene is characterized by one motional constant, $c_{3z^2-r^2}$.

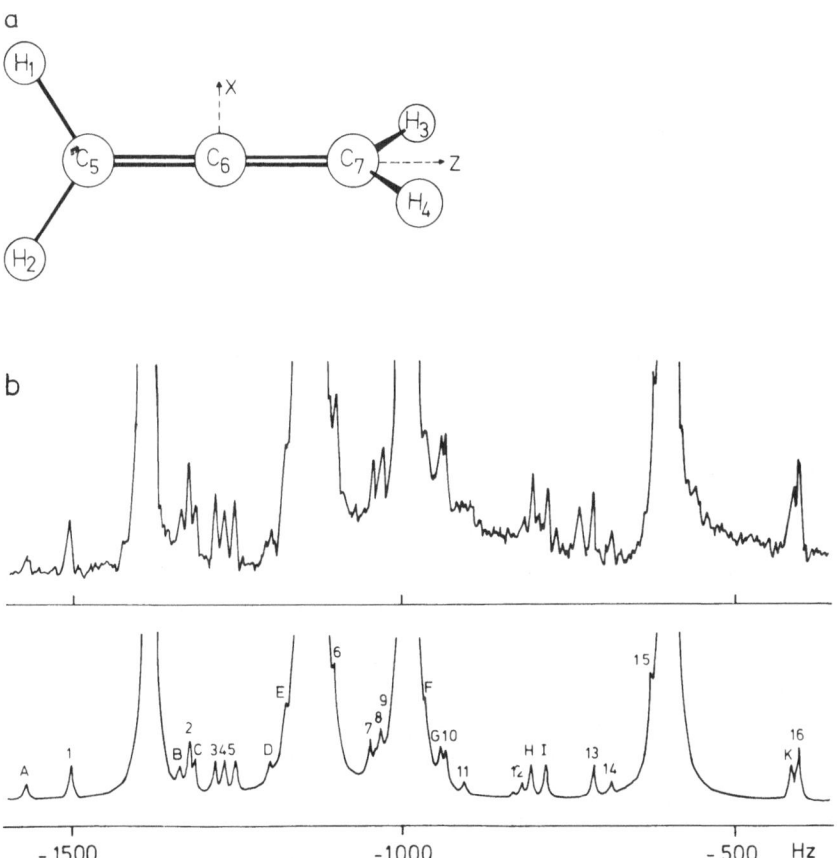

Figure 8. (a) Allene molecular model and notation.
(b) Proton resonance spectrum of allene in nematic 4.4′-hexyloxyazoxybenzene at high sensitivity (at 80°C). Upper trace: experimental spectrum; lower trace: computer simulated spectrum. Lines denoted by numbers are due to the molecules H$_2$^{13}C=C=CH$_2$. Lines denoted by letters are due to H$_2$ C=^{13} C=CH$_2$.

Limitations of the method and other results

At present the applicability of nematic liquid crystals as anisotropic solvents in nmr spectroscopy seems to be limited to relatively simple and symmetric molecules which do not have more than 8 protons [40, 44]. Beyond this limit the spectra become prohibitively complicated and extensive overlapping of the lines occurs. Thus, a simple molecule as cyclohexane yields only a very broad spectrum [44].

A·very promising way out of these difficulties has been proposed recently [44]. This method is based on the use of partially deuterated molecules. By a special double resonance technique (which makes use of double quantum transitions), a complete decoupling of the deuteron-proton coupling can be achieved [44] which leads to considerable simplified spectra. Using this double resonance technique, Meiboom and Snyder [44] succeeded in analyzing the proton nmr spectrum of cyclohexane. Obviously, quite complicated and large molecules could, in principle, be studied and analyzed with this technique if species that are deuterated at specific sites could be prepared.

Determination of stable molecular conformations

The liquid crystal is well suited to determine the stable conformations of molecules which may exist in different geometric isomers. Such information can be obtained in many cases without performing a total structure determination. Gazzard and Sheppard [53] determined the stable rotational isomer of 1,2,2,3-tetrachloropropane from the ratio of two proton-distances by comparing the experimental value with the corresponding value calculated by assuming different geometric isomers. Cocivera [54] showed by the nmr method that s-trioxane exists predominantly in the chair conformation at $80°C$. Many other examples can be found in the review by Diehl and Khetrapal [43].

Structure determination of non-rigid molecules

Care must be taken in analyzing the nmr spectra of oriented non-rigid molecules. The spectra are extremely sensitive to small changes in the molecular geometry because such changes affect both the bond distances and the order parameters. The spectra depend on the ratio of the life time of a metastable molecular configuration to the correlation time of the molecular tumbling motion. If the molecular tumbling is fast compared to the internal conversion between two molecular conformations, the direct coupling constants are weighted averages of the coupling constants of each conformation. In this case each conformation has its own motional constant. If the contrary holds, one has to calculate first the average bond distances and assume then an order parameter for this averaged geometry. This situation has been found to be valid for the cyclobutane molecule [40]. This molecule is rapidly flipping between two bent conformations.

The sensitivity of the direct coupling constants against small changes in the molecular geometry can also be exploited for the determination of barriers for internal conversions (or rotations). Diehl et al. [55] determined the barrier for the rotation of the CH_3-group in substituted toluenes. The magnitude of the barrier was estimated by comparing the motional constants and the relative molecular geometry obtained from the nmr spectrum with the corresponding values calculated by assuming different models for the form of the rotational barrier. It was shown that o-chlorotoluene assumes a staggered conformation with a barrier height of about 1.2 kcal/mole.

Information on barrier heights can in principle also be obtained by studying the temperature dependence of the nmr spectra. With the development of low melting nematogenes temperature dependent studies may now be performed between about $-50°C$ and $+200°C$. Thus the liquid crystal method seems to provide a very powerful tool for future studies of internal molecular conversions.

Optical active molecules

Due to the extreme sensitivity of the order parameters to small changes in the molecular geometry it is also possible to distinguish between the nmr spectra of the d- and l-isomers of optical active molecules [56]. The two isomers of a racemic mixture of trichloropropylene oxide (IV) yield different nmr spectra in a compensated nematic mixture of cholesteryl derivatives (Section I).

$$H_2 \diagdown \qquad \diagup CCl_3 \qquad\qquad\qquad\qquad\qquad\qquad (IV)$$
$$C \!-\!\!-\!\!- C$$
$$H_1 \diagup \quad O \quad \diagdown H_3$$

Such a compensated mixture provides an environment which exhibits both anisotropy and helicity. The two isomers have somewhat different order parameters in a solvent with a local screw sense and correspondingly exhibit different values of the direct coupling constants. In an ordinary nematic solvent (such as n-hexyloxyazoxybenzene) the two isomers give identical spectra. [56].

The direct coupling constants for one isomer I are:

$$D_{13}^I = +347.5; \; D_{23}^I = -14.0; \; D_{12}^I = -505.0 \; Hz.$$

The corresponding values for the other isomer II are:

$$D_{13}^{II} = +363.0; \; D_{23}^{II} = -15.5; \; D_{12}^{II} = -520.5 \; Hz.$$

Assuming that the orientation of the molecule is primarily determined by only one order parameter S_{zz} (which describes the average orientation of the axis z normal to the molecular plane) and assuming a value of $R_{12} \approx 1.65$ A one estimates $S_{zz}^I = -0.00475$ and $S_{zz}^{II} = -0.0049$.

2.2. Electron-Spin-Resonance Spectroscopy

2.2.1. Introduction

The application of liquid crystals as anisotropic solvents in the esr-spectroscopy has yielded less spectacular results than the application in high resolution nmr spectroscopy. However, the esr spectroscopy is a most important and versatile technique for investigations of the molecular properties of the liquid crystals [57, 58]. Stable free radicals (such as nitroxide radicals), chelates of transition metal ions (such as vanadyl chelates) [57, 58] and excited triplet states of aromatic molecules [59, 60] may be incorporated into liquid crystals as paramagnetic probes. The esr spectra of these probes provide information on the molecular arrangement, the dynamical state and the anisotropic intermolecular interactions in liquid crystals. This so called spin label technique is becoming of supreme importance for the investigation of artificial or biological membranes [15, 61, 62, 63, 64]. The spin label technique is however beyond the scope of this book and will not be considered in the following.

Up to the present, liquid crystalline solvents have been used (a) to determine the anisotropies of both the g-factor and the hyperfine interaction in ground state doublet (free) radicals [65] and (b) to study the dipole–dipole- and the electron exchange-interaction in species containing two unpaired electrons (biradicals and ground state triplets) [66, 67, 68]. More recently liquid crystals have been used extensively for studies of the anisotropic properties of photochemically excited triplet states in aromatic molecules and molecular complexes [60, 69].

2.2.2. Spin Hamiltonian for Free Radicals

The esr spectra of radicals with one unpaired electron can be accounted for by the Hamiltonian:

$$\mathcal{H} = \vec{H}_o \cdot g \cdot \vec{S} + \sum_i \vec{S} \cdot A_i \cdot \vec{I}_i \tag{10}$$

β is the Bohr magnetone, g is the electronic g-factor tensor and A is the tensor which accounts both for the isotropic and the anisotropic hyperfine interaction of the unpaired electron spin with the nucleus i. $\vec{S}$ and $\vec{I}_i$ are the operators of the electronic and the nuclear angular momentums, respectively. The vectors $\vec{S}$ and $\vec{I}_i$ are most conveniently expressed in terms of a space fixed coordinate system U V, W (Fig. 13).

It is convenient to orient the space-fixed coordinate system in such a way that the W-axis is directed parallel to the external magnetic field, $\vec{H}_o$. $\vec{H}_o$ provides a space-fixed quantization axis for the spin operators. The g- and the A-tensors are molecular properties which are defined in a molecular fixed principal axes system x, y, z. It is assumed in the following that the relative orientation of the coordinate systems U, V, W and x, y, z is described by the Eulerian angles ϕ, ψ, θ (Fig. 13). For a quantitative analysis of the esr spectra on the basis of the above Hamiltonian, $\mathcal{H}$, it is necessary to express $\mathcal{H}$ in the molecular fixed axes system. The corresponding transformation can be achieved by rotating the spin operators $\vec{S}$ and $\vec{I}_i$ about the Eulerian angles ϕ, ψ, θ. An extensive discussion of this transformation has been given by Glarum and Marshall [70, 71], by Falle and Luckhurst [65] and by Luckhurst [57].

Neglecting nonsecular terms[1], the transformed Hamiltonian may be expressed as [65, 70, 71, 72]

$$\mathcal{H}_{eff} = \beta H_o S_z \{g - (g - g_{zz})S_{zz} + 1/3 (g_{xx} - g_{yy}) (S_{xx} - S_{yy})\}$$
$$+ \sum_i S_z I_{i,z} \{a_i - (a_i - A_{zz,i})S_{zz} + 1/3 (A_{xx,i} - A_{yy,i}) (S_{xx} - S_{yy})\} \tag{11}$$

a_i and g denote the isotropic hyperfine coupling constants and the isotropic g-factor, respectively. These two parameters determine the line positions in isotropic media. They are given by one third of the traces of the corresponding tensors A and g $(a_i = 1/3 (A_{i,xx} + A_{i,yy} + A_{i,zz}); g = 1/3(g_{xx} + g_{yy} + g_{zz}))$.

The above equation is valid for radicals which have at least C_{2v} or D_{2h} symmetry, that is for which $S_{xy} = S_{xz} = S_{yz} = 0$.

For asymmetric molecules the Hamiltonian may be written in an analogous form if the principal axes system of the g- and the hyperfine tensor coincide with the principal axes system of the Saupe order matrix S [65].

---

[1] The non-secular terms lead to a small splitting of second order which cannot be observed in most cases [70].

2.2.3. Evaluation of Experiments

Together with the order parameters, Equ. (11) contains 8 unknowns (2 × 3 tensor components and 2 independent order parameters). The situation is considerably simplified if the molecule (and the interaction tensors A and g) exhibit cylinder symmetry. In this case $S_{xx} - S_{yy} = 0$ and the average radical orientation is described completely by only one order parameter S_{zz}. Moreover $A_{xx} = A_{yy}$ and $g_{xx} = g_{yy}$.

According to Equ. (11) the partial alignment of a radical in a liquid crystal causes only shifts in the line positions. In contrast to the situation in nmr spectroscopy no new lines appear as a consequence of the partial orientation. The components of the g- and A-tensor and the elements of the order matrix always enter in combination into Equ. (11). Consequently these parameters cannot be determined simultaneously from the line shifts caused by the partial alignment even in the simplest case of cylindrical molecules.

The following two methods have been proposed in order to solve this problem. All these methods are based on a comparison of calculated and experimental interaction parameters.

Determination of S_{zz} from splitting constants [65, 71]

The isotropic coupling constant a_i is determined by the so-called Fermi contact interaction between the unpaired electron and the nuclei [65]. The anisotropic contributions to the hyperfine coupling constants

$$b_i = 1/2(a_i - A_{zz,i}) \qquad (11a)$$

are determined by the direct dipole–dipole interaction between the magnetic moments of the electron and the nuclei, respectively. McConnell and Strathdee [73] calculated both the isotropic Fermi contact and the anisotropic dipole–dipole interactions of an unpaired 2pπ orbital (centered at a carbon atom i) with the ^{13}C nuclei (a (C_i) and b (C_i)) and with the protons (a (H_i) and b (H_i)) of the hydrocarbon skeleton. These authors derived simple relations between the coupling constants a_i or b_i and the so-called spin densities ρ_i of the unpaired 2pπ-orbital at the carbon nuclei, C_i, and at the protons, H_i. Using these relations, the coupling constants of organic free radicals can be easily calculated, provided the spin densities are known. In many cases the spin densities are known to a good accuracy from quantum mechanical calculations [74, 75, 77].

The following procedure for the analysis of the esr spectra can therefore be applied: first some of the anisotropic and (or) isotropic coupling constants are calculated using the McConnell and Strathdee relations. By comparing calculated coupling constants with measured line shifts, the order parameters, necessary to describe the orientation of the radical considered, are obtained. With the known order parameters the rest of the coupling constants and the g-factors (g_{xx}, g_{yy}, g_{zz}) are then obtained from the experimental line shifts. Falle and Luckhurst [65] applied this method to about 10 free radicals. In some cases these authors obtained valuable information on the geometry of the radicals. Moreover, the signs of the isotropic coupling constants and of the spin densities could be determined in these cases. For a detailed outline of the method and for examples of application the reader is referred to the original paper by Falle and Luckhurst [65]. An example of this method will be given below.

Determination of S_{zz} from the g-factor shift [72]

S_{zz} can in principle be determined from the shift in the g-factor caused by the partial alignment. According to Equ. (11) the experimental g-factor shift, Δg^{exp}, in going from an isotropic to a nematic solvent is given by

$$\Delta g^{exp} = (g - g_{zz}) \, S_{zz} \tag{11b}$$

Hall and Hardisson [75] and Stone [76] showed that to an accuracy of 20% an average value of $g_{zz} = 2.00238$ can be assumed for the z-component of the g-tensor of hydrocarbon radicals. For radicals with a threefold or higher symmetry axis S_{zz} can therefore be determined to an accuracy of 10% [72].

2.2.4. Example of Application

The effect of the partial orientation is demonstrated in Fig. 9 for the perinaphthenyl radical in azoxyanisol. This spectrum has been taken at the isotropic-nematic transition temperature [72] of the solution. Due to a small temperature gradient in the resonator the sample contained both isotropic and nematic regions on recording the spectrum. Fig. 9b gives therefore a superposition of the spectra of both the randomly and the partially oriented radicals.

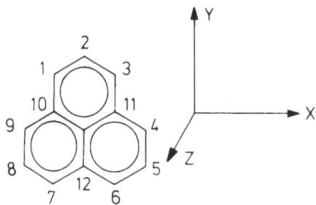

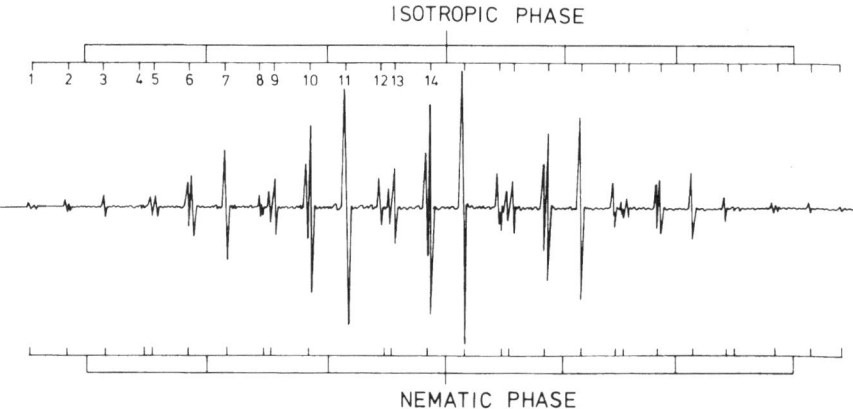

ISOTROPIC PHASE

NEMATIC PHASE

Figure 9. (a) Perinaphthenyl radical.
(b) Esr spectrum of perinaphthenyl radical in p-azoxyanisole at the isotropic–nematic transition temperature (135°C). The spectrum is a superposition of the spectra of randomly and partially oriented species. Note that the spectrum consists of seven quartets (relative intensity 1:3:3:1) with the relative intensities 1:6:15:20:15:6:1 (reprinted from Möbius et al. [72]).

The 28 line spectrum is composed of seven groups of 1:3:3:1 quartets with relative intensities 1:6:15:20:15:6:1. The septet splitting is caused by the protons at the positions 1,3,4,6,7 and 9 forming one group of equivalent protons. The quartet splitting is due to the protons at the 2, 5 and 8 position which form another equivalent group. The splitting of two adjacent quartet lines is given by

$$a^*(H_2) = a(H_2) + (a(H_2) - A_{zz}(H_2)) S_{zz}$$

and the splitting of two adjacent septet lines yields the value of $a^*(H_1)$. These values are given in Table 4. Fig. 9 and Table 4 show that the splitting of the quartet lines (that is $a^*(H_2)$) is increased whereas the splitting of the septet lines is decreased by the partial alignment of the radical in the liquid crystalline phase. Comparison of the line positions in the isotropic and in the nematic phase yields the shifts $\Delta a_i = (a_i - A_{i,zz})S_{zz}$ and $\Delta g = (g - g_{i,zz})S_{zz}$.

By recording the spectrum at very high spectrometer sensitivity, the additional lines caused by the hyperfine interaction of the unpaired electrons with the ^{13}C isotrops at natural ^{13}C abundance of 1.1% may also be observed. These lines yield the splittings $a^*(C_i)$ given in Table 4.

Table 4. Comparison of the coupling constants ($a_i^* = a_i + (a_i - A_{zz,i})S_{zz}$) and the g-factor ($g^* = g + (g - g_{zz})S_{zz}$) of the perinaphthenyl radical taken in an isotropic and in a nematic phase of p-azoxyanisol. (Values in Gauss from Möbius et al [72]).

	$a^*(H_1)$	$a^*(H_2)$	$a^*(C_1)$	$a^*(C_{2,10})$	$a^*(C_{13})$	g^*
Isotropic (140°C)	-6.277	1.830	9.744	-7.845	3.4(?)	2.002599
Nematic (117°C)	-6.262	2.010	6.496	-6.969	3.05(?)	2.002685

Determination of the order parameter of the perinaphthenyl radical

a) From the coupling constants [70]: McConnell and Strathdee [73] showed that the isotropic proton hyperfine coupling of the unpaired π-electron in perinaphthenyl is proportional to the spin density ρ_i of the electrons at the ith carbon atom: $a(H_i) = -27\rho_i$. According to Snyder and Amos [77] the spin density at the position 1 ($\rho_1 = 0.219$) is considerably larger than at positions 2 ($\rho_2 = 0.064$) or at position 10 ($\rho_{10} = 0.054$). Therefore the anisotropic hyperfine coupling of the $^{13}C_1$ nucleus, b (C_1), is to a good approximation directly proportional to the spin density ρ_1. The ratio of the isotropic coupling constant a (H_1) to the anisotropic coupling constant b (C_1) (cf Equ. 11a)) is therefore independent of the spin density (cf Ref. [70]).

$$b(C_1)/a(H_1) = 1.018 \pm 0.02 \text{ or } S_{zz} = \frac{\Delta a(C_1)}{2.036\, a(H_1)} \tag{12}$$

b) From the g-factor shift [72]: the order parameter S_{zz} can be calculated from the experimental g-factor shift g^{ex} (cf Equ. 11b) using a value of $g - g_{zz} = 6 \times 10^{-5}$.

The order parameters determined by the two different methods (a) and (b) agree within 10% [72]. Once the order parameter is known, the anisotropic proton and ^{13}C coupling constants in perinaphthenyl radical can be determined.

2.2.5. Other Applications

Haustein et al. [78] showed recently that the liquid crystal method can also be applied to study properties of electrolytically generated radicals.

Dinse et al. [79] succeeded in measuring ENDOR spectra in room-temperature nematogenes. The ENDOR technique enables one to study radicals of low symmetry. It offers also a method to determine nuclear quadrupole coupling constants in free radicals.

2.3. Triplet-esr-Spectroscopy of Partially Oriented Molecules

The esr-spectroscopy is one of the most important techniques to detect and study photochemically excited (paramagnetic) triplet states of aromatic molecules or of molecules of biological interest. By his technique, very small changes in the electron distribution caused by inter- and intramolecular interactions (e.g. substitution effects, solvent effects, complex formation) may be detected [69].

The triplet esr spectra are determined primarily by the dipole–dipole interaction of the two unpaired electrons. This interaction causes a splitting (the so called zero-field-splitting (ZFS)) of the threefold degenerate triplet state even in the absence of an external magnetic field H_o. The Hamiltonian for the triplet states is obtained by adding to Equ. (11) a term, H_{ZFS}, which takes care for the dipole–dipole interaction:

$$H_{ZFS} = -XS_x^2 - YS_y^2 - ZS_z^2 \tag{13}$$

X, Y and Z are the components of the so-called zero-field-splitting tensor, which is defined in the molecular fixed (principal) coordinate system x, y, z. For photochemically excited triplet states $H_{ZFS} \sim 0.1 \, cm^{-1}$.

The extreme sensitivity of the ZFS-components to small changes in the electron distribution provides the basis for the application of the triplet-esr spectroscopy as an analytical tool. A prerequisite for such applications is the assignment of the ZFS-components to the molecular axes. In the past this goal has been achieved by incorporating the molecules of interest in a suitable host single crystal [80] or by a magneto-photoselection technique [81].

As shown below, the assignment of the ZFS-parameters to the molecular axis can be made very easily by using liquid crystals as host to orient the triplet molecules. Since esr-spectra of excited triplet states can only be observed in rigid solvents ordered glasses are needed for such experiments. As noted in Section 1 uniaxial glasses can easily be prepared from the electric field oriented nematic mixtures of cholesteryl derivatives.

A typical, so-called $\Delta m = 1$ spectrum of randomly oriented triplet molecules is shown in Fig. 10. The shape of the spectrum is determined by the high anisotropy of the ZFS-splitting. The absorption spectrum (upper trace in Fig. 10) exhibits discontinuities in the intensity at six distinct field positions $H_x^{II,III}$, $H_y^{II,III}$ and $H_z^{II,III}$. Correspondingly, the first derivative esr spectrum exhibits sharp lines at these

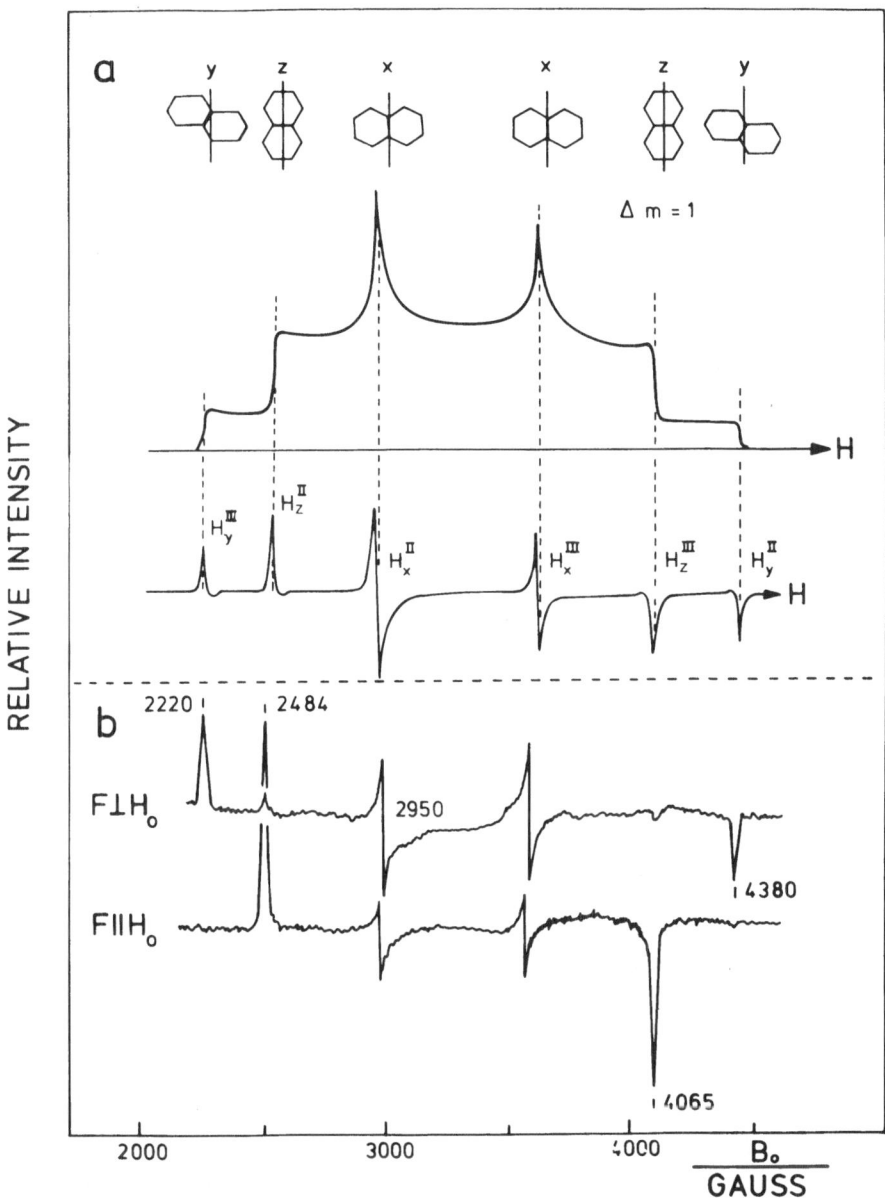

Figure 10. (a) Esr-spectrum of the lowest excited triplet state of randomly oriented naphthalene.
Upper trace: esr absorption spectrum.
Lower trace: first derivative esr spectrum.

Figure 10. (b) Esr-spectrum of naphthalene in an oriented glass (a compensated nematic mixture of cholesteryl chloride and cholesteryl laurate) taken at 77K.
Upper trace: liquid crystal optic axis $\vec{L}$ perpendicular to H_0.
Lower trace: liquid crystal optic axis $\vec{L}$ parallel to H_0. Note that $\vec{L}$ is parallel to the orienting electric field.

field positions (Fig. 10a, lower spectrum). It is a well known rule in triplet esr
spectroscopy that only those molecules contribute to the esr intensity at these
distinguished field positions, $H_{x_i}^{II,III}$, for which the magnetic field is parallel, or
nearly parallel, to a principal axis of the ZFS-tensor [82] (so-called canonical
orientations). To a first approximation the ZFS parameters X, Y, Z (= X_1, X_2, X_3)
may be expressed in terms of the resonance fields $H_{x_i}^{II,III}$ [83] (Equ. (36)). The
relative intensities at the fields positions $\underset{\rightarrow x_i}{H^{II,III}}$ are proportional to the number ΔN_i
of molecules for which the magnetic field $\vec{H}_o$ is (within a range of about $10°$)
parallel to the principal axis x_i of the ZFS-tensor [82]. In a liquid crystal (glass),
the number ΔN_i (and correspondingly the relative esr intensities) depends
1. on the orientation distribution of the solute in the liquid crystal and
2. on the orientation of the liquid crystal optic axis $\vec{L}$ relative to the external
    magnetic field direction $\vec{H}_o$.
    The effect of the partial alignment on the intensity distribution of triplet esr-
spectra is shown in Fig. 10b for naphthalene as guest. Spectra are shown for an
orientation of the liquid crystal optic axis $\vec{L}$ either perpendicular (upper trace;
$\vec{L} \perp \vec{H}_o$) or parallel (lower trace; $\vec{L} \parallel \vec{H}_o$) to the external magnetic field $\vec{H}_o$. Comparison of Figs. 10a and 10b shows that for $\vec{L} \parallel \vec{H}_o$ the intensity of the pair of
lines at $H_z^{II,III}$ is greatly increased, while the intensity of the lines at $H_x^{II,III}$ is
greatly reduced by the partial alignment of naphthalene in the liquid crystal matrix.
The contrary holds if the optic axis is oriented perpendicular to the external
magnetic field direction ($\vec{L} \perp \vec{H}_o$). Now, it is known from optical polarization
experiments (cf Sections 3 and 4) that an elongated molecule such as naphthalene
orients with its long axis, z, preferentially parallel to the optic axis of the liquid
crystal. Thus the z-lines (the position of which is determined by the ZFS parameter
Z) are due to those molecules for which $\vec{H}_o$ is parallel to the long molecular axis.
Accordingly, the tensor component Z has to be assigned to the long molecular axis.

A quantitative analysis of the relative intensities is only possible on the basis of a very
elaborate computer simulation of the spectra [60]. Such an analysis yields an estimation of the
orientation distribution function of the solute molecules in the liquid crystal matrix [59]. As
will be shown in section 4, it is physically reasonable to assume an exponential distribution
function, which (for molecules of D_{2h}- and C_{2v} symmetry) is given by

$$\rho (\theta_x, \theta_y, \theta_z) \propto \exp \{c_x \cos^2\theta_x + c_y \cos^2\theta_y + c_z \cos^2\theta_z\},$$

Here $\cos \theta_x$, $\cos \theta_y$ and $\cos \theta_z$ denote the direction cosines of the liquid crystal optic axis, $\vec{L}$,
in the solute molecular fixed coordinate system (x, y, z). By comparing the computer-simulated
and the experimental intensities at the canonical field positions ($H_x^{II,III}$, $H_y^{II,III}$, $H_z^{II,III}$) one
obtains the coefficients c_x, c_y, c_z, by trial and error method. The coefficients c_x, c_y, c_z can
also be determined to a very good approximation by comparing the maximum intensities of the
first derivatives lines at the canonical orientations [60, 84]. The elements of the order matrix
can easily be calculated using the above distribution function. The triplet esr spectroscopy
provides thus a most useful method for the measurement of the solute orientation in liquid
crystals. This method can be applied to a large class of solute molecules.
Liquid crystals have been extensively used in recent esr spectroscopic studies of charge-
transfer complexes [60 and 69].

3. Applications of Liquid Crystals in Optical Spectroscopy

3.1. Absorption and Fluorescence Polarization Spectroscopy

3.1.1. Introduction

Optical polarization experiments provide most important information both on the symmetry and the vibrationally induced mixing of excited molecular states. Moreover, polarized spectra may be used for identifying hidden optical transition. Polarization measurements require that the molecules are at least partially oriented with respect to an external axis. Several methods have been applied in the past to prepare systems of homogeneously oriented solute molecules for this purpose. In the classical experiment by McClure [85] this goal has been achieved by incorporating the molecules in suitable host single crystals. Another method, that is the orientation of the molecules by strong electric fields [86, 87], is restricted to very polar solute molecules. Most frequently used is the method of photoselection, where the solute alignment is achieved by irradiating the solute molecules (embedded in a rigid organic glass) with polarized light [88, 89, 90, 91, 92]. This method yields, however, only relative polarization directions. An excellent review of this technique has been given recently by Dörr [93].

The most straightforward way to produce partially oriented solute molecules is to orient them in anisotropic solvents. Good solute orientation can be achieved in stretched polymer sheets [94, 95, 96]. Homogeneously oriented nematic liquid crystals are perfectly clear and are thus excellently suited as anisotropic solvents for optical polarization experiments. Moreover, the liquid crystal method allows the performance of polarization experiments in fluid media. This unique feature of the liquid crystal method can be exploited for polarization studies of metastable molecular species (e.g. excited complexes) formed by a diffusion-controlled process. The ordered glasses produced by rapid cooling of a uniformly aligned nematic phase can be used for phosphorescence polarization experiments.

As an additional benefit absolute values of the order matrix elements of solute molecules may be determined from polarized absorption spectra. This optical method can be applied to a large class of organic solute molecules.

3.1.2. Principle of the Method

Two types of polarization experiments are conceivable:

Linear Dichroism Experiments

Polarized absorption (or fluorescence) spectra are taken in uniformly aligned nematic phases. The orientation of the optical transition moment in the molecular fixed coordinate system of the solute can be determined from the change in the optical density of the solute produced by the partial solute alignment [7].

Circular Dichroism Experiments

A strong circular dichroism (CD) is observed in the absorption region of molecules embedded in cholesteric phases. The sign of the CD yields information on the relative orientation of the optical transitions in a very easy way.

Solvents. For most purposes the liquid crystal should be transparent (non-absorbing) in the ultraviolet wavelength region. As described in Section 1, the prerequisite of transparency is to some extent met by the mixtures of cholesteryl derivatives. By recrystallization from ethanol and subsequently from dioxane, the molar extinction coefficients of cholesteryl chloride and cholesteryl laurate at 230 nm may be reduced to $\epsilon \sim 3$ l mole^{-1} cm^{-1}. For absorption polarization experiments in the wavelength region of the lowest wavelength transition of large aromatic molecules (such as tetracene) the aromatic nematogenes of type I may be used as solvents [20]. Some liquid crystals are non-fluorescent and are thus also suited for fluorescence polarization studies [99, 100].

Experimental Procedure

a) *Linear Dichroism.* The liquid crystal method of polarization measurements is extremely simple if the aromatic nematic compounds of type I can be used. In this case the orientation of the solutions is effected between flat quartz plates as described in Section 1. A drawback to this method is that only small light paths of about 50 μm are possible [20].

For the experiments with the compensated nematic mixtures of the cholesteryl derivatives one has to prepare cells which allow the application of electric fields of the order of 10,000 V/cm. Depending on the solubility of the solute the light path of the cells may be varied between several μm and 1 cm. Electrode distances between 0.4 and 1.0 cm are possible. For measurements of the order parameters, the light path of the cells should be made as small as possible in order to minimize depolarization effects caused by thermal fluctuations of the solvent order [101]. Depolarization effects may also be minimized by placing polarizers on either side of the cell [7]. A schematic representation of the experimental arrangement for the absorption and polarization experiments is shown in Fig. 11. The preparation of ordered

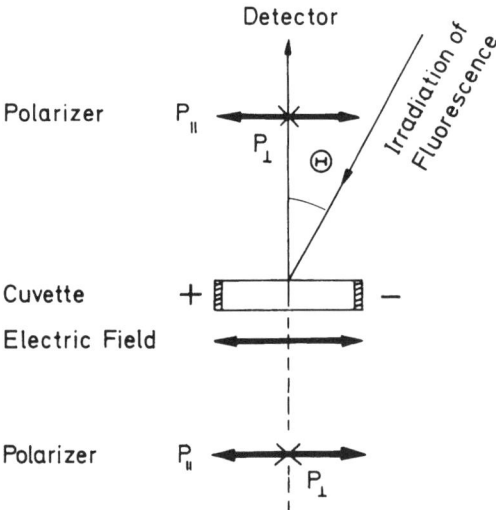

Figure 11. Schematic representation of the experimental arrangement for the absorption and fluorescence polarization experiments. The excitation of the fluorescence may be effected with unpolarized or polarized light.

glasses for the phosphorescence polarization experiments has been described in Section 1.

A convenient measure of the degree of absorption polarization, N_A, is the ratio of the optical densities (OD)

$$N_A = \frac{OD_\parallel}{OD_\perp} \tag{14}$$

$OD_\parallel$ and $OD_\perp$ are the optical densities obtained if the polarization direction of the measuring light is oriented parallel and perpendicular to the optic axis $\vec{L}$ (or the electric field $\vec{E}$), respectively. In the same way the degree of fluorescence polarization is defined by the corresponding ratio of the fluorescence intensities $I_\parallel$ and $I_\perp$ ($N_F = I_\parallel/I_\perp$). For quantitative measurements the absorption and fluorescence spectra have to be corrected for the self-polarization of the instruments [7].

b) *Circular Dichroism.* The samples are simply prepared by sandwiching the cholesteric solutions between flat quartz plates separated by spacers of variable thickness (to about 50 μm). By application of a small pressure the cholesteric phase orients in such a way that the helix (or optic axis) axis is directed perpendicular to the glass surface. The samples are oriented in the spectro-polarimeter with the helix axis parallel to the beam of the measuring light.

3.1.3. Interpretation of the Polarized Spectra

Linear Dichroism

In the following we derive relations between the degree of polarization and the solute orientation. We assume that the optical axis $\vec{L}$ (director) of the homogeneously oriented liquid crystal is directed parallel to the axis W of a space fixed coordinate system U, V, W (Fig. 12). The average orientation of the solute molecular symmetry

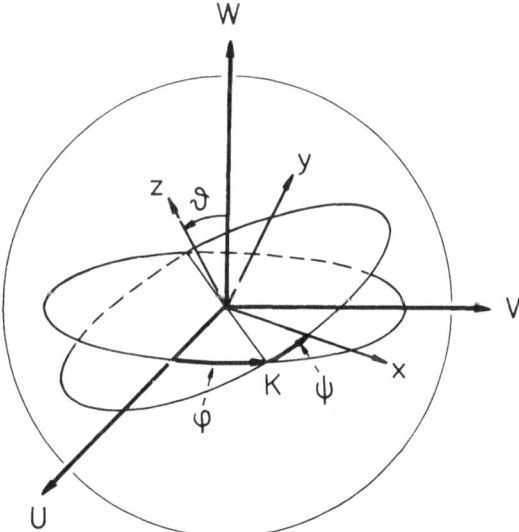

Figure 12. Orientation of a solute molecule in the liquid crystal matrix. U, V, W is a space fixed axes system. The optical axis $\vec{L}$ of the homogeneously oriented liquid crystal is assumed to be parallel to W. The solute molecule fixed coordinate system x, y, z is rotated by the Eulerian angles ϕ, ψ, θ with respect to the space fixed axis system U, V, W.

axes $(x, y, z \equiv x_1, x_2, x_3)$ with respect to the optical axis W may be expressed in terms of the order matrix.

The absorption spectra of aromatic molecules consist of a number of absorption bands corresponding to different electronic transitions. Each electronic band is composed of a superposition of vibrational bands v centered about the wavelength λ_v, where each band v corresponds to a transition between two vibrational levels of different electronic states. The probability W_{kl} of an optical transition between two electronic states k and l is proportional to the square of the projection of the transition moment vector $\vec{M}_{kl}$ on the polarization direction $\vec{e}$ of the measuring light $(W_{kl} \propto |\vec{M}_{kl} \cdot \vec{e}|^2)$. For reasons of simplicity we consider in the following symmetrical molecules of D_{2h}- or C_{2v}-symmetry, for which each vibrational band (at λ_v) is polarized parallel to any of the symmetry axes $(x, y, z \equiv x_1, x_2, x_3)$. Each vibrational band is therefore characterized by only one component $M^v_{x_i}$ of the transition moment [102]. The following relation holds between the molar extinction coefficient $\epsilon^v_{x_i}$ and the transition moment $M^v_{x_i}$ ($\tilde{\nu}$ is the wavenumber in cm^{-1}):

$$\int_{-\infty}^{+\infty} \epsilon^v_{x_i} (\tilde{\nu} - \tilde{\nu}_v) \, d\tilde{\nu} = 1.1 \ 10^{38} \ n \ \tilde{\nu}_v |M^v_{x_i}|^2 \tag{15}$$

Due to the double refraction of liquid crystals the refractive index n depends on the angle between the polarization direction of the measuring light and the optic axis $\vec{L}$ (parallel to W). For cholesteryl derivatives the double refraction can be neglected $(n_\parallel - n_\perp \sim 0.03)$ [109]. For normal nematogenes of type I the double refraction is not negligible.

The elements ϵ_{x_i} of the molar extinction coefficient tensor are defined in the solute molecular fixed axes system, whereas the polarization direction of the absorbed and the emitted light is defined in the space-fixed axes system $(U, V, W \equiv u_1, u_2, u_3)$. The optical density at the wavelength λ for light polarized parallel to an external axis u_i is given by

$$OD_{u_i} (\lambda) = cd \sum_v \epsilon^v_{u_i} (\lambda) = \epsilon_{u_i} (\lambda) \ cd \tag{16}$$

where c and d are the concentration and the length of the light path in the cuvette, respectively. The summation is extended over all vibrational bands v contributing to the absorption at the wavelength λ. The molar extinction coefficients in the direction of the axis u_i, $\epsilon^v_{u_i}$, are given by [7]

$$\epsilon^v_{u_i} (\phi, \psi, \theta) = \sum_k A_{u_i x_k} \cdot \epsilon^v_{x_k} \cdot A_{x_k u_i} \tag{17}$$

$A_{u_i x_k}$ are the elements of the well known rotation matrix which transforms the solute molecular axes system into the laboratory system by rotation about the Eulerian angles ϕ, ψ, θ. ϵ_{u_i} depends therefore on the orientation of the solute molecules with respect to the laboratory fixed coordinate system. The average molecular extinction coefficient of an ensemble of solute molecules, $<\epsilon_{u_i}>$, in a direction u_i is obtained by averaging $\epsilon_{u_i} (\phi, \psi, \theta)$ over the anisotropic orientation distribution of the guest molecules as follows [7]:

$$\epsilon_\perp = \epsilon_u = \epsilon_v = \bar{\epsilon} - 1/3 \ (\epsilon_x S_{xx} + \epsilon_y S_{yy} + \epsilon_z S_{zz})$$

$$\epsilon_\parallel = \epsilon_w = \bar{\epsilon} + 2/3 \ (\epsilon_x S_{xx} + \epsilon_y S_{yy} + \epsilon_z S_{zz}) \tag{18}$$

where $\bar{\epsilon} = 1/3 \ (\epsilon_x + \epsilon_y + \epsilon_z)$.

By inserting these equations in Equ. (14) the following equation is obtained for the degree of absorption polarization:

$$N_A = \frac{OD_{\parallel}}{OD_{\perp}} = \frac{3\bar{\epsilon} + 2 (\epsilon_x S_{xx} + \epsilon_y S_{yy} + \epsilon_z S_{zz})}{3\bar{\epsilon} - (\epsilon_x S_{xx} + \epsilon_y S_{yy} + \epsilon_z S_{zz})} \qquad (19)$$

For liquid crystals with non-negligible double refraction the corrected degree of polarization is given by

$$N_A^{corr} = \frac{n_{\perp}}{n_{\parallel}} N_A$$

A similar relation as Equ. (19) holds for the degree of fluorescence polarization. The molar extinction coefficients are replaced by the sum of the fluorescence intensities $f_v (\lambda_v - \lambda)$ of the individual vibrational bands v contributing to the fluorescence spectrum at the wavelength λ. These intensities depend on the cube of the refractive index [103]:

$$\int f_v (\bar{\nu} - \bar{\nu}_v) \, d\bar{\nu} \propto n^3 \, (\vec{M}^v)^2 \qquad (20)$$

Thus the birefringence of the liquid crystal is much more critical in the fluorescence polarization studies. Measurements of the order parameter of excited molecules are possible if the rotational relaxation time of the solute is small compared with the lifetime of the emitting state. The birefringence of the liquid crystal is also of critical importance if one wishes to determine the solvent order from the absorption polarization spectrum of the solvent in its nematic phase. The complications arise from the strong refractive index dispersion in the wavelength regions of the solvent absorption bands. Saupe and Maier [104] have treated this case. The same difficulties arise of course if the solute and the solvent absorption overlap.

According to Equ. (19) the degree of polarization for molecules of C_{2v} or D_{2h}-symmetry depends on five unknown parameters, namely $S_{xx} - S_{yy}$; S_{zz}; ϵ_x; ϵ_y; ϵ_z. One of the extinction coefficients can be determined from the optical density measured in an isotropic solvent, where $OD = c \cdot d \cdot \bar{\epsilon}$. For planar aromatic molecules the number of unknowns reduces to three since the $\pi - \pi^*$-transitions of these molecules are polarized parallel to the molecular plane.

In many cases the 0-0-transitions are well separated from higher vibrational transitions and are therefore "purely" polarized parallel to an in-plane symmetry axis. In these cases the order parameters of solute molecules can be determined from the degree of polarization of two perpendicularly polarized o-o-transitions using the following equation:

$$S_{x_i x_i} = \frac{N_{x_i} - 1}{N_{x_i} + 2} \qquad (21)$$

Provided the values of the order matrix elements are known, the ratio of the extinction coefficients, $(\epsilon_x / \epsilon_z)$, in the directions of the in-plane molecular axes can be determined by $(\epsilon_y = 0)$:

$$\frac{\epsilon_x}{\epsilon_z} = - \frac{1 + 2 S_{zz} - N (1 - S_{zz})}{1 + 2 S_{xx} - N (1 - S_{xx})} \qquad (22)$$

This ratio of the extinction coefficients yields quantitative information on the vibrationally induced mixing of excited states.

According to the above discussion a quantitative evaluation of the absorption spectra in oriented solvents is much simpler than an analysis of magnetic resonance spectra of partially oriented molecules. However, this is only valid if the axes of preferred solute orientation (or the principal axes of the order matrix) coincide with the directions of the optical transition moments. The situation may be much more complicated if molecules of lower than C_{2v} symmetry are considered. In these cases the principal axes system of the order matrix and the directions of the 0-0-transitions moments do not coincide. An example for such a molecule is chrysene [7]. In these cases the liquid crystal method yields only the relative polarization directions. A way out of these difficulties could be a systematic study of substituted molecules. It is well known that the substituents such as methyl groups do not shift the polarization direction appreciably but such groups may have large effects on the average orientation of the molecule in the liquid crystal.

Circular Dichroism

Cholesteric liquid crystals can be understood to be composed of stacked, two-dimensional nematic layers while the optic axis $\vec{L}$ of these nematic layers rotates parallel to the helix axis. Solute molecules orient within these nematic layers qualitatively in the same way as in ordinary nematic liquids. Thus the guest molecules embedded in cholesteric phases exhibit the same helical arrangement as the solvent (host) molecules. The very strong circular dichroism (CD) [98, 105] exhibited by cholesteric solutions in the absorption region of the solute is a consequence of this helical arrangement of the solute [98]. This "structural" CD can be understood on the basis of theory on light propagation in cholesteric phases developed by De Vries [106] and by Mauguin [107]. The essential predictions of this theory are:

1. Any planar wave entering the cholesteric crystal in a direction parallel to the helix axis is split up into two elliptically polarized components which are rotating in opposite directions with somewhat different velocities.

2. The minor axis of one wave coincides with the maximum of the other one.

3. The axes of the ellipses are parallel and perpendicular, respectively, to the local optic axes $\vec{L}$ of the nematic layers. Consider an elongated solute molecule possessing an optical transition which is polarized parallel to its long axis. In each nematic layer this molecule is oriented with its long axis preferentially parallel to the optic axis $\vec{L}$. Consequently, its transition moment is also oriented preferentially parallel to this local optic axis. Obviously such a solution will absorb more energy from the elliptically polarized component with the maximum amplitude parallel to the director $\vec{L}$ than from the wave with the minor axis parallel to $\vec{L}$. Thus a linearly polarized wave entering such a cholesteric liquid crystalline solution will leave as an elliptically polarized wave. The ellipticity Θ (ratio of the minor to the maximum axis) of the outcoming wave depends obviously on the difference of the molar extinction coefficients for the two elliptically polarized components. Now, any linearly polarized wave entering the cholesteric solution can be considered as being composed of two oppositely rotating circularly polarized waves of equal intensity. Obviously these two oppositely rotating waves must suffer different absorptions to yield an elliptically polarized outcoming wave. Accordingly the cholesteric solution exhibits circular dichroism in the absorption region of the guest molecule. The ellipticities of the two elliptically polarized waves built up in the cholesteric phase depend critically on the pitch of the helix. Thus the circular dichroism is a

sensitive function of the pitch. Depending on the ratio, $\lambda^* = \lambda/\lambda_{max}$, of the wavelength, λ, of the measuring light to the wavelength of maximum reflectivity, λ_{max}, of the cholesteric phase, two limiting cases may be evaluated quantitatively [98].
1. If λ^* is large compared to the relative birefringence $\alpha = (n_\parallel - n_\perp)/(n_\parallel + n_\perp)$, of the solvent, the ellipticity Θ is to a good approximation given by:

$$\tan \Theta = 0.25 \frac{\alpha}{\lambda^*} \tanh \left| \frac{2.3}{2} (\epsilon_\parallel - \epsilon_\perp) \, c \, d \right| \tag{23}$$

where $\parallel$ and $\perp$ refer to the directions parallel and perpendicular to the local optic axis $\vec{L}$, respectively.
2. For very large values of the pitch, that is if $\lambda^* \ll \alpha$, the coefficient in front of the hyperbolic tangent (tanh) in Equ. (23) must be replaced by λ^*/α.

These two equations are valid for a cholesteric phase which forms a right-handed screw with respect to the direction of light propagation. The same expressions with negative signs are obtained for left-handed cholesteric phases.

The appearance of the difference $\epsilon_\parallel - \epsilon_\perp$ in the argument of the hyperbolic tangent shows that the sign of the circular dichroism depends upon whether the transition is polarized parallel or perpendicular to the long molecular axis of the solute molecules. The circular dichroism spectrum therefore yields in a very simple way information on the relative polarization directions of optical transitions.

It is well known that the CD is always accompanied by an optical rotation dispersion (Cotton effect). Such a Cotton effect has also been observed [108]. It is clear that a helically arranged system of emitting molecules should exhibit also a fluorescence circular dichroism. Such an effect has been reported very recently [110].

3.1.4. Examples of Application

Anthracene

The absorption polarization spectrum of oriented anthracene is presented to about 250 nm in Fig. 13a. Anthracene orients with its long axis preferentially parallel to the optical axis of the liquid crystal. It therefore follows from Fig. 13 that the ^1L_a-transition with the 0-0-vibrational band at 26,600 cm^{-1} is preferentially polarized along the short in-plane molecular axis, while the 0-0-band at 39,800 cm^{-1}, corresponding to the ^1B_b-transition, is polarized parallel to the long axis of anthracene. This result is in complete agreement with theoretical predictions and with the results of the photoselection technique [90].

The degree of absorption polarization N and the ratio ϵ_z/ϵ_x (Fig. 13b) reveals a slight vibrational structure in the range of the longest wavelength (^1L_a-) absorption. It is well known from photoselection experiments [92] that this vibrational structure in the degree of absorption polarization is caused by a vibrationally induced mixing of the higher excited ^1B_b-state (which is polarized parallel to the long molecular axis) to the ^1L_a-state. A consequence of this mixing is the appearance of two series of vibrational bands: one series being polarized in the same direction as the 0-0-band of the ^1L_a-transition and a second series being polarized in the direction of the ^1B_b 0-0-transition. The second series is caused by the vibrationally induced mixing. Accordingly, the first series yields minima of the ratio $OD_\parallel/OD_\perp$, while the second series yields maxima of N.

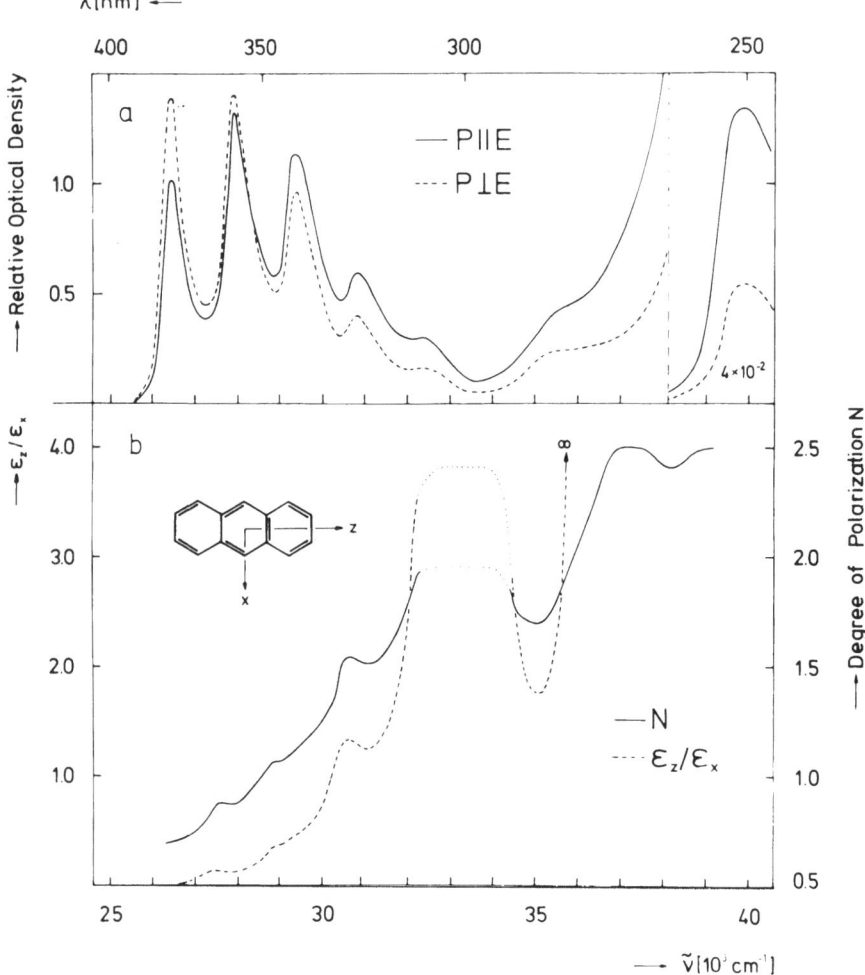

Figure 13. Absorption polarization spectrum of anthracene (reproduced from Sackmann and Möhwald [7].

(a) Absorption spectrum in a 1.85:1 by weight mixture of cholesteryl chloride and cholesteryl laurate ordered at $T_{nem} = 30°C$ by an electric field of 10,000 V/cm. Solid line (———): spectrum taken with polarizers oriented parallel to the electric field (P‖E). Broken line (— — —): polarizer perpendicular to $\vec{E}$ (P⊥E). $\tilde{\nu} = 25,000 - 38,000$ cm^{-1} : concentration c = 4 × 10^{-3} m/ltr., $\tilde{\nu} > 38,000$ cm^{-1} c = 1.6 × 10^{-4} m/ltr., respectively.

(b) Solid line: the degree of polarization defined as the ratio of the optical densities N = OD$_{‖}$/OD$_{⊥}$. Broken line: ratio of the molar extinction coefficients ϵ_z/ϵ_x along the in-plane molecular axes x and z.

The ^1B_b 0-0-transition at 32,800 cm^{-1} is very strong and it is therefore reasonable to assume that the transition at this wavenumber is to a good approximation polarized purely parallel to the long molecular axis of anthracene. From the value of N = 2.5 at the maximum of the ^1B_b-band (at 250 nm) one obtains $S_{zz} = + 0.33$.

The 0-0-band at the long wavelength and of the ^1L_a transition is polarized "purely" parallel to the short in-plane axis and yields $S_{xx} = -0.08$. With these values of the order parameter, the ratio ϵ_z/ϵ_x has been calculated using Equ. (22).

The degree of absorption polarization and the ratio of the extinction coefficient ϵ_z and ϵ_x exhibits a sharp increase starting at about 29,000 cm^{-1}. This indicates that a weak transition, namely the ^1L_b-transition, is buried below the much stronger ^1L_a-band at wavelengths above 29,000 cm^{-1}. A comparison of the absorption polarization spectrum, the fluorescence polarization spectrum and the circular dichroism spectrum shows that the O-O-vibrational band of the ^1L_b transition is located at about 29,000 cm^{-1} [7]. This example demonstrates the usefulness of the liquid crystal method to localize weak hidden optical transitions.

Pyrene-2-carbonic-methyl ester

An example of a circular dichroism spectrum is presented in Fig. 14 together with the corresponding absorption polarization spectrum. Pyrene-2-carbonic-methylester exhibits bands at 263 nm, 282 nm, 336 nm and 395 nm. Clearly, the molecule orients with its long axis preferentially parallel to the optic axis (or the electric field $\vec{E}$). Accordingly, the bands at 263 and 336 nm are polarized parallel to the long molecular axis (z), while the bands at 282 and 395 nm are polarized perpendicular to the z-axis. Fig. 14 shows very clearly that the CD changes its sign when the degree of polarization changes from a value $N < 1$ to a value $N > 1$ in accordance with Equ. (23). It could not be decided from the absorption spectrum taken in an isotropic solvent if the weak band at 282 nm originates from a new electronic transition or if it is due to a higher vibrational band of the transition centered at 336 nm. The polarized spectra and the CD spectrum show very clearly that the weak band is due to a new electronic transition.

Other Examples

As mentioned above, an advantage of primary importance of the liquid crystal method is that polarization measurements may be performed in fluid media. This advantage has been exploited recently for fluorescence polarization studies of metastable molecular complexes formed in the excited state (e.g. excimers [111] and exciplexes [99]).

3.2. Infrared Spectroscopy

It is clear that the liquid crystal method can also be applied to perform polarization studies in the infrared (ir) spectroscopy. However, the ir-spectroscopy in liquid crystals is hampered by the background absorption of the solvent. In favorable cases the solute ir bands are located in wavelength regions where the solvent absorption is not significant. p-(p'-ethoxybenzoxy)phenyl butyl carbonate (Id) is only weakly

$$CH_3(CH_2)_3-O-\overset{O}{\overset{\|}{C}}-O-\!\!\!\bigcirc\!\!\!-\overset{O}{\overset{\|}{C}}-O-\!\!\!\bigcirc\!\!\!-OC_2H_5 \qquad\qquad (Id)$$

absorbing (absorbance smaller than OD $\sim$ 0.15) in the wavelength region between 1900 and 2800 cm^{-1}. Terminal $C \equiv O$ stretching bands usually occur between 1900

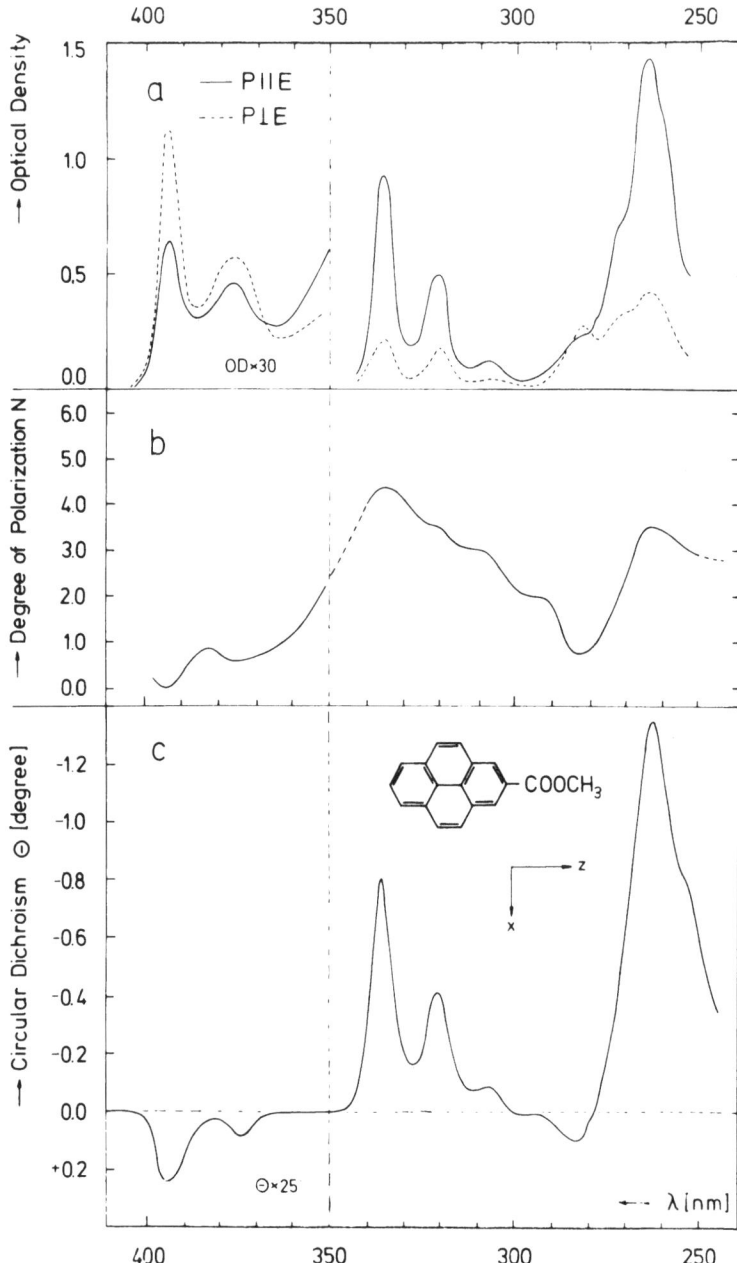

Figure 14. (a) Absorption polarization spectrum of pyrene-2-carbonic methyl ester (PCME). Solvent: 1.85:1 by weight mixture of cholesteryl chloride and cholesteryl laurate ($T_{nem} \approx 30°C$). Concentrations: $10^{-3}M$ and $3.3. \, 10^{-2} M$, respectively. The spectrum P∥E was taken with the linear polarizers P oriented parallel to the electric field $\vec{E}$ (or parallel to the optix axis $\vec{L}$).
(b) Degree of polarization N (cf Equ. (14)) as determined from the absorption polarization spectrum. N yields the order parameters $S_{xx} = -0.162$; $S_{yy} = -0.345$ and $S_{zz} = +0.515$.
(c) Circular dichroism spectrum. Solvent: cholesteric mixture of 55.5 mole percent cholesteryl nonanoate and 44.5 percent cholesteryl chloride at 30°C. Concentration $c = 5 \times 10^{-3}$ mole/ltr.; solvent pitch: $\lambda_{max} = 900$ nm; sample thickness: $d = 20 \, \mu m$. Spectrum taken with a Cary 60 spectropolarimeter.

and 2200 cm^{-1}. This particular liquid crystal is thus well suited to determine the orientation of $C \equiv O$ stretching transitions.

Homogeneous orientation of this liquid crystal is easily achieved between two barium fluoride plates which have been rubbed in one direction and which are separated by teflon spacers (of about 30 μm thickness). As an example the polarized infrared spectrum of $Re_2(CO)_{10}$ is shown in Fig. 15 in the wavelength region of the $C \equiv O$ stretching band [112]. The $Re_2(CO)_{10}$ molecules orient with their long molecular axis z (Fig. 15a) preferentially parallel to the optic axis $\vec{L}$. Thus the two bands at 2070 cm^{-1} and at 1970 cm^{-1} are polarized parallel to the long molecular axis, whereas the intense band at 2010 cm^{-1} is polarized perpendicular to the z axis. The $Me_2(CO)_{10}$ molecules exhibit D_{4d}-symmetry and, correspondingly, have three allowed $C \equiv O$-stretching normal modes: two modes of B_2 symmetry and one of E_1 symmetry. According to group theory, the B_2 vibration transforms as the long molecular axis, z, of $Me_2(CO)_{10}$ under the molecular symmetry operations and

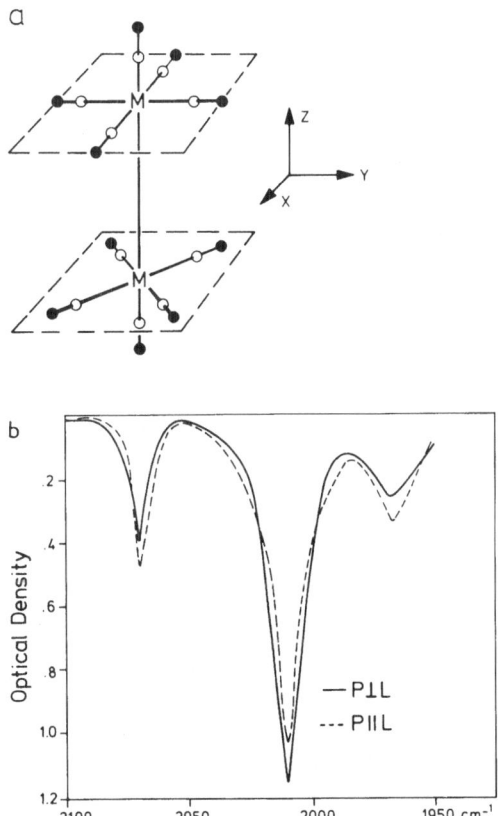

Figure 15. (a) Structure of $Re_2(CO)_{10}$, M: metal atom; (o) carbon atoms; (•) oxygen atoms. (b) Polarized infrared spectrum of 2.7 × 10^{-2}M $Re_2(CO)_{10}$ in p-(pethoxybenzoxy)phenyl butyl carbonate. The spectra are corrected for background absorption (reprinted from Ceasar et al. [20]).

is thus polarized parallel to this axis. In the same way, group theory predicts that
the E_1 vibration is polarized perpendicular to the z-axis.

The above example shows that the assignment of the vibrational bands is
achieved in a most simple way using the liquid crystal method. The same assignment
of the vibrational bands of $Re_2(CO)_{10}$ has been obtained earlier by Flitcroft et al.
[113] on the basis of a much more complicated analysis of the intensity and the
calculated force constants.

A most interesting version of the liquid crystal method for the determination of the ir
polarization has been proposed by Kelker et al. [22]. This method makes use of homeotropically
ordered nematic phases: first the spectrum is recorded (without polarizer) in a homeotropically
oriented nematic layer, then the spectrum is taken after heating the sample above the isotropic
state. The primary advantage of this method is that it does not need polarizers. Information on
the polarization direction is obtained by comparing the optical densities measured in the randomly
and in the uniformly ordered phase, respectively. Consider a rod-shaped molecule (rod axis z).
Then $S_{xx} = S_{yy} = -1/2\,S_{zz}$. The effective molar extinction coefficient for unpolarized light
that is traversing the homeotropically ordered system in a direction parallel to the optic axis is
$\epsilon_{eff} = 1/2\,(\epsilon_u + \epsilon_v)$ (Fig. 12). For a band that is polarized parallel to the rod axis the ratio of
the optical density measured in the ordered phase to the density measured in the isotropic state
is (Equ. (18)):

$$N\,(\epsilon_z) = \frac{OD_{ordered}}{OD_{isotropic}} = 1 - S_{zz}$$

and for a band that is polarized perpendicularly to the rod axis

$$N\,(\epsilon_x) = 1 + \frac{1}{2}\,S_{zz}.$$

Since S_{zz} is a positive quantity (rod axis preferentially parallel to $\vec{L}$) it is $N\,(\epsilon_z) < 1$ and
$N\,(\epsilon_x) > 1$.

4. Anisotropic Intermolecular Interaction Potential and Solute Orientation

Both the degree of order in liquid crystals and the average orientation of guest
molecules in liquid crystals are closely related to the anisotropy of the intermolecular
forces. The measurements of the solute or the solvent order are therefore most
important in order to test theoretical models of the forces acting between nonspherical molecules. The use of nematic phases as model systems for the investigation
of anisotropic intermolecular interaction potentials is another important scientific
application of liquid crystals.

The starting point for a theory of the anisotropic intermolecular interaction in
liquid crystals is the Maier and Saupe theory [114, 115, 116, 118]. This theory is based
on the assumption that the intermolecular interaction potential in nematic liquid
crystals is determined primarily by London dispersion forces. The effective anisotropic potential U of a molecule ℓ in the anisotropic dispersion field generated by
its oriented neighbors s is calculated by averaging the pair potential between two
molecules ℓ and s over all orientations of the "solvent" molecules s and over all

distances between ℓ and s. For rod-like molecules the anisotropic contribution to this average potential is given by:

$$W_\ell^{an}(\theta_\ell) = -\frac{A S_s}{2 V^2}(3\cos^2\theta_\ell - 1) \tag{24}$$

A is a constant which depends on the nature of the liquid crystal, V is the molar volume of the liquid crystal and S_s is the order of the nematic phase. θ_ℓ is the angle between the long axis of the molecule ℓ and the optic axis of the liquid crystal. In the approximation of the so-called mean (molecular) field approach [117, 118] the average orientation of molecule in the liquid crystal is completely determined by the above anisotropic potential. Applying simple **Boltzmann** statistics, the average orientation of the long axis of a molecule can then be expressed as

$$S_{zz} = \int_0^\pi\int_0^{2\pi}(3\cos^2\theta_\ell - 1)e^{-W_\ell^{an}/kT}\sin\theta_\ell d\theta_\ell d\phi_\ell / \int_0^\pi\int_0^{2\pi} e^{-W_\ell^{an}/kT}\sin\theta_\ell d\theta_\ell d\phi_\ell \tag{25}$$

The Maier and Saupe theory successfully accounts for the observed temperature dependence of the order of nematic phases and correctly predicts the existence of a first order transition at a temperature

$$T = \frac{A}{4.542\, kV_t^2}, \tag{26}$$

where V_t is the molar volume at the transition point. Although the quantitative agreement between the theoretically predicted and the observed temperature dependence of the solvent order is not satisfactory in many cases [120] the **Maier and Saupe** theory should provide a basis for a theoretical understanding of the anisotropic solute–solvent interaction in liquid crystalline solutions. Accordingly, the average solute orientation should be related to the anisotropic solute solvent interaction potential W_ℓ^{an} by a relation identical to Equ. (24). As noted above, quantitative information on the average solute orientation is often obtained as a by-product of the application of liquid crystals in spectroscopic studies. Large paramagnetic probes such as nitroxide radicals [120] or vanadyl-chelates [70, 119, 84] as well as optical probes [7] may yield information on the order of the liquid crystal matrix.

Consider a uniaxial liquid crystal. The anisotropic interaction potential of a solute molecule in the mean anisotrop field of the solvent may be expressed as [121]:

$$W_\ell^{an} = \sum_{ik\ell} b_{ik\ell}\cos^i\theta_1 \cos^k\theta_2 \cos^\ell\theta_3 \tag{27}$$

$\cos\theta_1$, $\cos\theta_2$ and $\cos\theta_3$ denote the direction cosines characterizing the orientation of the solute molecular fixed axes x_i with respect to the optic axis $\vec{L}$.

In a first approximation W_ℓ^{an} may be approximated by[2]

$$W_\ell^{an} = -a_0 - a_1\cos^2\theta_1 - a_2\cos^2\theta_2 - a_3\cos^2\theta_3 \tag{28}$$

---

[2] The neglection of terms linear in $\cos\theta_i$ excludes polar forces.

The absolute values of the coefficients a_i depend both on the nature of the solute and the solvent. However, the ratios a_i/a_j are only functions of the physical properties of the guest molecules. By a systematic study of the orientation of different solute molecules in different solvents it should, in principle, be possible to correlate the formally introduced coefficients a_i with physical properties of the molecules. The results of some of such studies are summarized in the following.

The contribution of the permanent electric dipole moments to the intermolecular interaction potential has been studied by Saupe and Nehring using nmr spectroscopy [116, 121]. These authors compared the order parameters of solute molecules with strong permanent dipole moments (such as 1, 2, 3-trichlorobenzene) in nematic solvents with electric dipole moments either oriented parallel (positive dielectric anisotropy) or perpendicular (negative anisotropy) to the long molecular (optic) axes[3]. If the anisotropic interaction potential and correspondingly the order parameters would be determined considerably by the forces between the permanent dipole moments of the solute and the solvent, one would expect a considerable difference in the order parameters as measured in the two solvents. No such effect could be observed for chloride [116] or fluoride [121] substituted benzenes. The coefficients a_i and the order parameters $S_{x_i x_i}$ agree closely for the two different nematic solvents. It can thus be concluded that the force between permanent electric dipoles does not contribute considerably to the anisotropic intermolecular interaction potential.

Saupe and Nehring [121] concluded from their measurement of the order parameters of a number of fluorobenzenes that
1. Each C—H or C—F bond contributes independently to the intermolecular potential (or to the coefficient a_i of Equ. (28)).
2. The interactions potentials of the C—H- and C—F-bonds with the solvent depend only on the angles between the bond axes and the solvent optic axis.
Accordingly, the coefficients a_i of Equ. (28) for substituted benzenes may be written as

$$a_i = a_B^i + B \sum_\rho \cos^2 \alpha_\rho^i \tag{29}$$

In this equation a_B is the coefficient for benzene and α_ρ^i is the angle between the bond axis of bond ρ and the solute fixed axis x_i. The above results of Saupe and Nehring [121] demonstrate the importance of σ-bonds for the solute orientation that is for the anisotropy of the solute-solvent interaction.

The additive contribution of the individual bonds to the anisotropic potential draws attention to the rule of the additivity of bond polarizabilities [122]. This strongly suggests that the average orientation of the substituted benzenes is directly related to their principal polarizabilities and that the anisotropic solute— solvent interaction is determined by London dispersion forces. Considering dispersion forces (in dipolar approximation) one obtains the following expression for the

---

[3] 4,4'-hexyloxyazoxybenzene was a convenient solvent with negative dielectric anisotropy. A 3:2 mixture of 4-heptanoyloxy- and 4-hexanoyloxy-4'-ethoxy azobenzene may be used as solvent with positive dielectric anisotropy.

anisotropic potential of a solute ℓ in the anisotropic environment of the solvent s on the basis of the molecular statistical theory [115, 117, 123]

$$W_\ell^{an}(\theta, \phi) = Q\{\alpha_{xx}\sin^2\theta\cos^2\phi + \alpha_{yy}\sin^2\theta\sin^2\phi + \alpha_{zz}\cos^2\theta\} \qquad (30)$$

θ and ϕ are the polar angles that characterize the orientation of the liquid crystal optic axis in the solute molecular fixed coordinate system. α_{xx}, α_{yy} and α_{zz} are the principal polarizabilities of the solute. The constant Q is primarily a function of the solvent properties.

By inserting the above expression into Equ. (25); one obtains a theoretical relation between the order parameters S_{xx}, S_{yy} and S_{zz} of a solute and its principal polarizabilities α_{xx}, α_{yy} and α_{zz}. This relationship is shown graphically in Fig. 16.

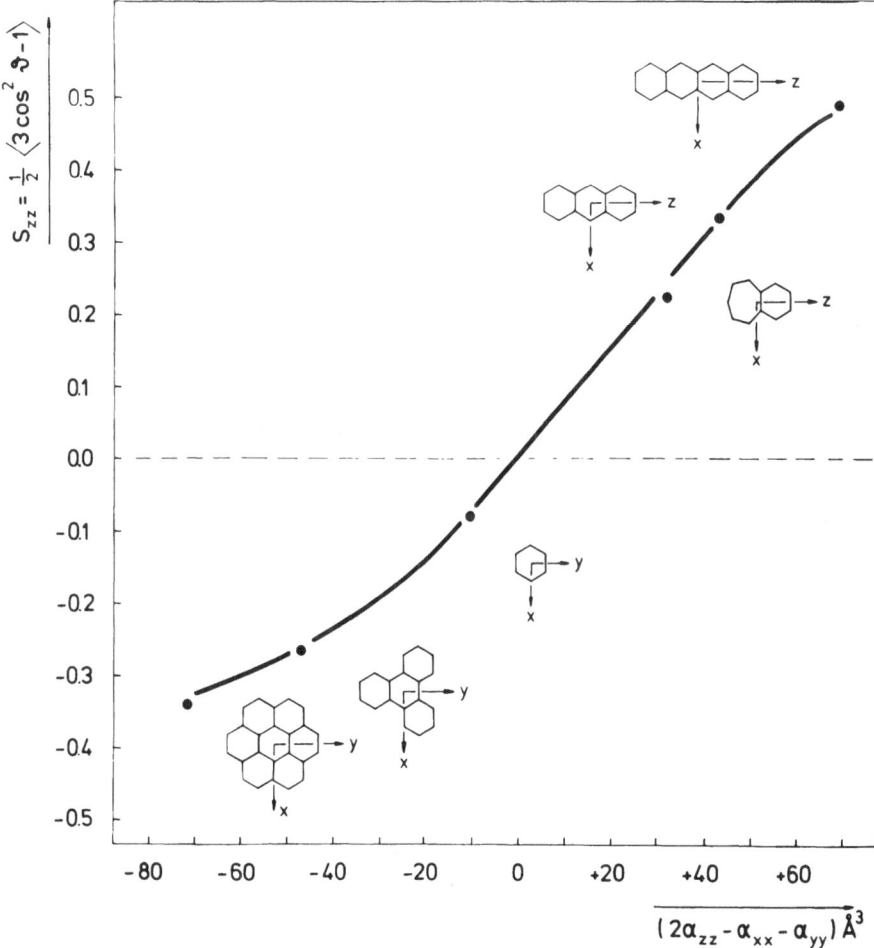

Figure 16. Relationship between the average orientations, S_{zz}, of the z-axes of some aromatic solute molecules and their principal polarizabilities $\alpha_{x_ix_i}$. Solid curve: Theoretical relation calculated from Equ. (30) for Q = 0.115 kT. Dots (•): experimental values of S_{zz} as observed in a compensated nematic, 1.8:1 by weight mixture of cholesteryl chloride and cholesteryl laurate for 30°C.

For a constant value of Q = 0.115 kT the order parameter S_{zz} has been plotted (solid curve in Fig. 16) as a function of the difference

$$\Delta\alpha = 2\,\alpha_{zz} - \alpha_{xx} - \alpha_{yy}$$

between the principal solute polarizabilities. For comparison in Fig. 16 are given also experimental order parameters for some aromatic molecules in a compensated nematic mixture of cholesteryl chloride and cholesteryl laurate. The principal polarizabilities have been taken from the literature [123]. The value of Q = 0.115 kT has been chosen in order to fit the calculated and the experimental S_{zz} value of anthracene. Using this value of Q, excellent agreement between calculated and experimental order parameter S_{zz} is also obtained for all other (unsubstituted) aromatic molecules considered in Fig. 16. In the same way the order parameters S_{zz} and S_{yy} characterizing the average orientation of the other symmetry axes x and y of the aromatic molecules in Fig. 16 may also be calculated to a good approximation using the above value of Q [123]. The result in Fig. 16 shows:
1. The average solute orientation of aromatic molecules in liquid crystals is determined primarily by anisotropic London dispersion forces.
2. The principal polarizabilities of aromatic molecules can be determined to a good approximation from the order parameters characterizing the orientation of the molecular symmetry axes in a nematic liquid. For this purpose the constant Q characteristic for the liquid crystal matrix must be determined [123] from the order parameters of a solute with known values of the principal polarizabilities.
Attempts have also been made to correlate the average orientation of solute molecules with their molecular dimensions [123, 124]. Such correlations should provide information on the importance of anisotropic repulsion forces for the solute orientation. Up to the present repulsvie forces have only been considered as important for theoretical reasons [125]. It has been shown recently [60] that the anisotropic repulsion potential may be represented by an expression similar to Equ. (30). The polarizabilities $\alpha_{x_i x_i}$ are replaced by the Van der Waal's dimensions, ℓ_{x_i}, in the directions of the molecular symmetry axes. By assuming that the solute orientation is determined primarily by repulsive forces one expects a correlation of the form $S_{zz} \propto 2\ell_{zz} - \ell_{xx} - \ell_{yy}$. According to Fig. 17 such a correlation holds for many elongated aromatic molecules. A similar relationship has also been found for substituted benzenes [124]. Thus it seems difficult to decide between dispersion and repulsive forces on the bases of the average solute orientation alone.

Judged from the above experimental results the anisotropic dispersion interaction seems to account successfully for the average solute orientation at least for aromatic solute molecules. It should be emphasized, however, that the applicability of the dipolar approximation adopted in the simple Maier and Saupe theory is questionable. This approximation has been made responsible for the failure of this simple theory to account in a quantitative way for the observed temperature dependence of the order in nematic phases [120]. In an attempt to overcome these difficulties a number of authors improved the Maier and Saupe theory.
1. Humphries et al. [120] extended the theory by including higher than dipolar terms in the average pair interaction potential. For this purpose the anisotropic

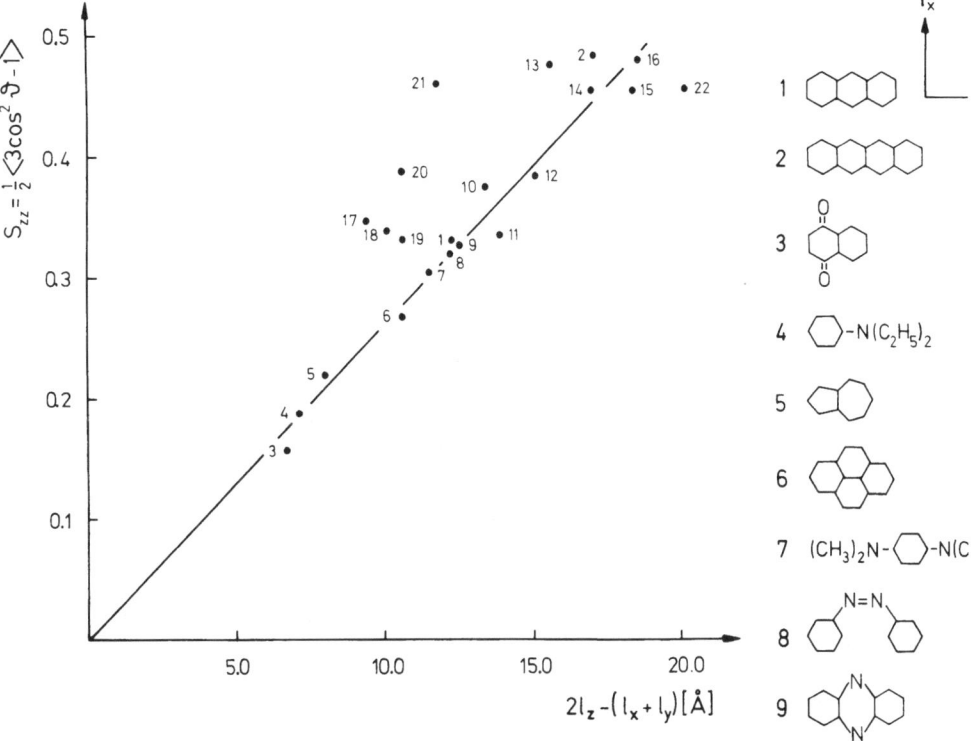

Figure 17. Relation between the average orientation S_{zz} of the long axes of elongated aromatic molecules Van der Waals lengths in the directions of the molecular symmetry axes. Note that molecules with highly S_{zz}-values than expected from their dimensions according to the linear relationship $S_{zz} \propto 2\,\ell_z - \ell_x - \ell_y$.

energy of a rod-like molecule in the field generated by the solvent s is expanded in spherical harmonics:

$$U(\cos \theta_\varrho) = \sum_L \bar{u}_L \, \overline{P_L(\cos \theta_s)} \, P_L(\cos \theta_\varrho) \qquad (31)$$

P_L are legendre polynomial of order L and θ_ϱ is the angle between the cylinder axis of the molecules and the optic axis. $\bar{u}_L$ are the expansion coefficients. The first term of this expansion is obviously identical with the dispersion potential of Equ. (30) since $S_s \propto \overline{P_L(\cos \theta_s)}$.

By fitting the observed and the calculated temperature dependencies of the order in 4,4' dimethoxyazoxybenzene Humphries et al. [120] obtained the following potential

$$U(\cos \theta_\varrho) = -C \{\bar{P}_2 P_2(\cos \theta_\varrho) - 0.184 \, \bar{P}_4 P_4(\cos \theta_\varrho)\} \qquad (32)$$

2. A severe drawback of the mean field approximation is the complete neglection of short range order effects (i.e. correlations between the orientation of adjacent molecules). In order to allow for short range order effects Chandrashekar and Madhusudana [126] and Wulf [125] tried to generalize the Maier and Saupe theory by taking into account rotational and translational correlations between adjacent molecules. Chandrashekar and Madhusudana consider the liquid crystal as being composed of clusters of perfectly oriented molecules. The statistical theory based on this model yields fair agreement between calculated and experimental temperature dependencies of the magnetic birefringence of nematic azoxyanisole. These examples show that liquid crystals may be of considerable interest as model systems for investigations of short range order effects in liquids.

5. Application of Liquid Crystals in Gas Liquid Chromatography

5.1. Basic Principles

According to the preceding Section 4, the anisotropic potential energy of a guest molecule in a liquid crystal is a function of its principal polarizabilities and consequently of its geometry. The heats of mixing (or the activity coefficients) of two geometric isomers (such as m- and p-xylene) may differ considerably in an anisotropic solvent although these isomers have identical activity coefficients in an isotropic solvent. The sensitive dependence of the activity coefficients on the molecular geometry provides the basis for the applicability of liquid crystals as stationary phases in gas–liquid chromatography (G–LC). The usefulness of liquid crystals as substrates has been reported for the first time by Kelker [131] and by Dewar and Schröder [127]. These authors showed that meta- and para-disubstituted benzenes may be separated much better by using liquid crystals as stationary phase than with conventional column-packing materials. Moreover, the gas–liquid chromatography is of particular interest for investigation of both the thermodynamic properties of liquid crystals and the specifity of solute–solvent interactions [128]. An important advantage of the G–LC method is that measurements can be performed with very small solute concentrations in order to minimize the disturbance of the liquid crystal by the solute. Moreover a wide variety of solute molecules can be studied rapidly and with high accuracy.

An extension of the Maier and Saupe theory to two component systems provides a basis for the theoretical interpretation of the solute activity coefficients in liquid crystals. Consider for this purpose a two-component system of rod-like molecules containing the mole fractions x of solute (ℓ) and $(1 - x)$ of solvent (s). By generalising Equ. (24) one obtains for the average anisotropic dispersion energy of the solute in the mixed solute—solvent system [129]

$$\overline{W_\ell^{an}} = (1 - x) A_{\ell s} S_s S_\ell + x A_{\ell\ell} S_\ell^2 \tag{33}$$

The first sum accounts for the anisotropic solute—solvent interaction, while the second term characterizes the anisotropic contribution to the solute—solute interactions. S_ℓ and S_s denote the solute and the solvent order, respectively. The coefficients $A_{\ell s}$ and $A_{\ell\ell}$ are negative quantities and are proportional to the coefficient A/V^2 (cf Equ. (24)). An analogous expression is valid for the solvent molecules. Accordingly, the total average orientational energy of the mixed solute—solvent system is given by [129]

$$\overline{U^{an}} = N_L \left\{ \frac{(1 - x)^2}{2} A_{ss} S_s^2 + x(1 - x) A_{s\ell} S_s S_\ell + 1/2 x^2 A_{\ell\ell} S_\ell^2 \right\} \tag{34}$$

The orientational contribution to the entropy, $S_{\ell(s)}^{an}$, of a solute (or a solvent) molecule, ℓ (or s), is determined by the orientational distribution function

$$F_{(\ell)s} = \frac{1}{Z_{\ell(s)}} \exp\{-W_{\ell(s)}^{an}(\theta, \phi)/kT\}. \tag{35}$$

$Z_{\ell(s)}$ is a normalization factor

$$Z_{\ell(s)} = \int_0^\pi \int_0^{2\pi} \exp\{-W_{\ell(s)}^{an}(\theta, \phi)/kT\} \sin\theta \, d\theta \, d\phi \tag{36}$$

one obtains

$$S_{\ell(s)}^{an} = -k \int F_{\ell(s)}(\theta, \phi) \ln F_{\ell(s)}(\theta, \phi) \sin\theta \, d\theta \, d\phi = \frac{\overline{W_{\ell(s)}^{an}}}{T} + k \ln Z_{\ell(s)} \tag{37}$$

The orientational part of the Helmholtz free energy is given by

$$A^{an} - \overline{U^{an}} - R T (x \ln Z_\ell + (1 - x) \ln Z_s) \tag{38}$$

A most convenient measure for the solute—solvent interaction is the solute activity coefficient γ. The activity coefficient γ_ℓ^∞ at infinite dilution is directly related to the retention volume V_g (or the retention time t_g[4]) by the equation [130, 132]

$$\gamma_\ell^\infty = \frac{RT}{V_g M_s p_\ell^o} \tag{39}$$

$$t_g = V_g/F,$$

where F is the volume that passes the column per unit time.

---

[4] The retention time is related to the retention volume by

In this equation, M_s is the molecular weight of the solvent and p_ϱ^o is the vapor
pressure of the pure solute. Provided the volume change caused upon mixing is
small, the orientational contribution to the activity coefficient, $\gamma_\varrho^{\infty, an}$, can be
calculated from the change in the Helmholtz free energy of mixing, ΔA^{an}, according
to

$$RT \, \ell n \gamma_\varrho^{\infty, an} = -\delta \Delta A^{an} / \delta x |_{x \to o} \tag{40}$$

For solutes which are not mesomorphic at temperatures where the G–LC experiments are performed it is $S_\varrho = 0$. In this case Equ. (40) leads to the following simple
equation for the orientational (anisotropic) contribution to the solute activity
coefficient

$$\ell n \gamma_\varrho^{\infty, an} = +1/2 \frac{A}{V^2 kT} S_s^2 - \ell n \, Z_\varrho \tag{41}$$

The coefficient $A_{s\varrho}$ of Equ. (34) has been replaced by $-A/2V^2$. According to this
equation, the activity coefficient is composed of two contributions: a solute
independent positive contribution and a second (negative) term, $\ell n Z_\varrho$, which depends
on the properties of the solute. The first term is a result of the heat of mixing,
while the second contribution accounts for the entropy of mixing. Equ. (41) leads
to the following general rules:
1. In general, the solvent-dependent first part of $\gamma_\varrho^{\infty, an}$ exceeds the entropy term
$\ell n \, Z_\varrho$. Accordingly, the activity coefficient increases (and the retention volume V_g
decreases) abruptly on going from the isotropic to the nematic (or smectic) state of
the liquid crystal. The Maier and Saupe theory predicts that at the nematic to
isotropic transition point, T_i, the first term in Equ. (41) should be about equal for
all nematogenes ($A/k \, T_i \, V_i^2 \approx 4.542$; $S_s \sim 0.445$ [114, 115, 129]). Accordingly the
change, $\Delta \gamma^\infty$, in the activity coefficient at the nematic-to-isotropic transition
should be solvent-independent ($\Delta \ell n \, \gamma^\infty \sim 0.4$). This prediction has been verified
experimentally by Chow and Martire [128].
2. In the temperature range of the stability of the mesophase the orientational
activity coefficient increases with decreasing temperature. This prediction has also
been verified experimentally [128].
3. The activity coefficients increase with increasing solvent order S_s. Such a
behavior has also been found experimentally [134].
4. The solute dependent contribution to γ_ϱ^∞ in Equ. (41), $\ell n Z_\varrho$, increases with
increasing anisotropy of the principal solute polarizabilities. This is easily verified
by inserting the potential of Equ. (30) in Equ. (36). For a cylindrical molecule
($\alpha_{xx} = \alpha_{yy} = \alpha_\perp$; $\alpha_{zz} = \alpha_\parallel$) one obtains

$$Z_\varrho \propto e^{a' \alpha_\perp} \sum_{n=o}^{\infty} \frac{(a')^n (\alpha_\parallel - \alpha_\perp)^n}{(2n + 1) \, n!} \tag{42}$$

According to Equ. (41) the anisotropic contribution to the retention volume V_g is
directly proportional to Z_ϱ. Therefore Equ. (42) predicts that the retention volume
is approximately proportional to the difference $\alpha_\parallel - \alpha_\perp$ of the principal polarizabilities (or proportional to the difference $\ell_\parallel - \ell_\perp$ of the molecular dimensions).
        Consider the case of the xylene isomers in order to test this rule: according to

Fig. 18 it is V_g (*ortho*) $> V_r$ (*para*) $> V_g$ (*meta*). Considering the anisotropies of
the polarizabilities alone one would expect, however, V_g (*para*) $> V_g$ (*ortho*) $>$
V_g (*meta*). The reason for this discrepancy is the neglection of the vapor pressure. The
m- and the p-isomers have about equal vapor pressures p_ℓ^o (p^o(m) = 350 torr at
$100°C$; p^o(p) = 353 torr at $100°C$), whereas o-xylene has a considerable lower
value of p_ℓ^o (P^o(o) = 300 torr at $100°C$). Thus by considering the vapor pressure,
rule 4 predicts the right order for the retention volumes of xylene isomers.

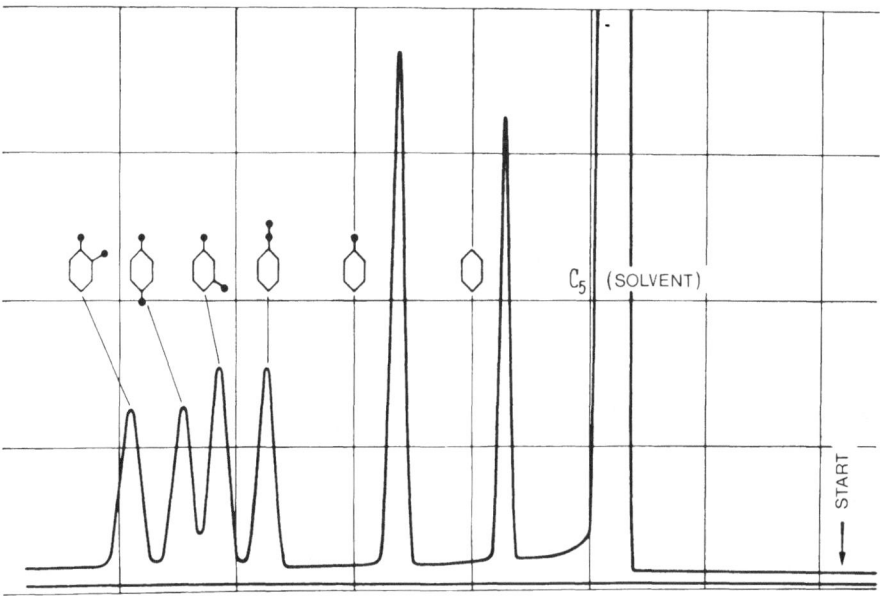

Figure 18. Chromatographic separation of xylene[1], ethylbenzene, toluene and benzene using
nematic p,p′-azoxyphenetole as stationary phase (values taken from Ref. [132]).

5.2. Examples of Application

For the preparation of the columns the liquid crystalline melts are simply
deposited on porous packing material or on capillary tubes [128, 131]. It is
important to maintain a uniform temperature over the whole column during the
measurement.

A considerable number of liquid crystalline solvents have been investigated
(Table 5). These studies showed that a necessary condition for a good separation
power of the liquid crystal is a large persistence range of the mesophase [131]. The
separation of *meta*- and *para*-xylene presents a convenient test for the effectiveness
of a column-packing material. The separation of these xylene isomers is a classical
problem in G–LC which is also of considerable technical importance. The two
isomers have about equal retention volumes in most ordinary stationary phases.

Fig. 18 shows, however, that excellent separation is achieved if one uses nematic p,p'-azoxyphenetol. The so-called separation factor

$$\alpha = V_g \, (meta)/V_g(para)$$

for azoxyphenetol is (at 140°C) α = 0.91. For Bentone 34, a commercially available clay with the best separation power of the known nonmesomorphic substrates one observes α = 1.13.

Fig. 19 shows the temperature dependence of the retention times of p- and m-xylene in 4,4'-dihexyloxyazoxybenzene. Good separation is observed both in the nematic (125 to 80°C) and in the smectic phase (T < 80°C). At the isotropic-to-nematic transition the retention time decreases in a small temperature interval (125–120°C). Such a behavior is expected according to rule 1 and has been observed in many cases [134, 128]. It should be emphasized that the drop in V_g at the isotropic-to-nematic transition takes place over a rather broad temperature range (that is 5°C), rather than sharply. This has been attributed to pretransitional effects [134].

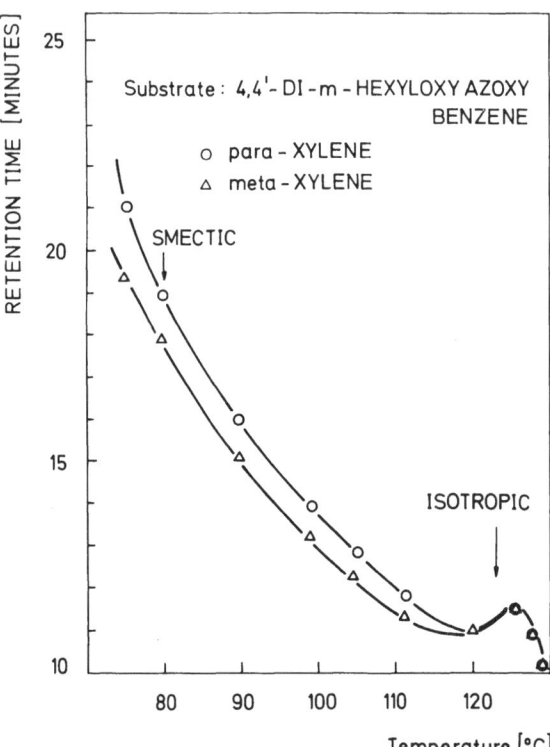

Figure 19. Temperature dependencies of the retention times of para- and meta-xylene in 4,4'-di-hexyloxyazoxybenzene. At the isotropic-to-nematic transition (at about 125°C) the retention times decrease abruptly in a temperature range of about 5°C. A good separation of the isomers is only possible in the liquid crystalline states of the substrate, where the activity coefficient (and thus the retention time) is determined primarily by the anisotropic solute–solvent interaction (from Ref. 127).

A further example of technical importance is the separation of *meta-* and *para-*divinylbenzene (DVB). The best separation factor obtained with conventional packing materials is $\alpha = V_g(meta)/V_g(para) = 0.91$. For di-n-hexyloxyazoxybenzene, a considerably higher separation factor of $\alpha = 0.77$ ($V_g(meta) = 1198$ ml/min; $V_g(para) = 1556$ ml/min), was obtained at 96°C. In accordance with rule 4 the *p*-compound has a higher value of the retention time than *m*-DVB.

For the separation of solutes with high boiling points Kelker et al. [135] synthesized cinnamic acid derivatives which may be used up to about 360°C. These liquid crystals are well suited for the separation of fatty acids.

As noted in Section 2 the *d-* and *ℓ*-isomers of optically active molecules exhibit somewhat different order parameters in liquid crystals possessing a local screw sense (such as cholesteryl derivatives). Accordingly *d-* and *ℓ*-isomers should separate on cholesteric substrates. For 3,3,3-trichloropropylene oxide the difference in the order parameter is of the two isomers is $\Delta S_{zz} = 0.00015$ (cf Section 2) for a compensated nematic mixture of cholesteryl derivatives. On the basis of Equ. (42) one would expect a separation factor of $\alpha \sim 1.002$. Up to the present the separation of optically active isomers on cholesteric substrates has not been achieved.

5.3. Thermodynamic Application of G–LC

The gas-liquid chromatography is a convenient technique for studying the thermodynamic properties of liquid crystals and liquid crystalline solutions. The basis for such applications is the following relation between the activity coefficient γ_ℓ^∞ and the partial molar excess free energy G_ℓ^E of the solute at infinite dilution

$$\ln \gamma_\ell^\infty = \frac{\Delta G_\ell^E}{RT} = \frac{\Delta H_\ell^E}{RT} - \frac{\Delta S_\ell^E}{R} \tag{43}$$

H_ℓ^E and S_ℓ^E are, respectively, the partial molar excess enthalpy and entropy. By plotting γ_ℓ^∞ as a function of the reciprocal temperature both H_ℓ^E and S_ℓ^E can be determined. Provided the molar enthalpy of vaporization, ΔH_ℓ^{vap}, of the solute is known, the partial molar enthalpies (ΔH_ℓ^{sol}) and entropies (ΔS_ℓ^{sol}) of solution may be obtained according to

$$\Delta H_\ell^{sol} = H_\ell^E - \Delta H_\ell^{vap}$$

$$\Delta S_\ell^{sol} = S_\ell^E - \Delta H_\ell^{vap}/T$$

Chow and Martire [128] determined these quantities for a large number of organic compounds in both the isotropic and nematic state of 4,4′-dimethoxyazoxybenzene and p-azoxyanisole.

6. Application of Liquid Crystals in Chemistry

6.1. Chemical Reactions in Ordered Fluids

An intriguing possible application of liquid crystals is their use as anisotropic solvent to study the influence of the partial orientation on the reaction rate of

Table 5. Examples of chromatographic separation with liquid crystalline substrates[a]

Stationary phase	Temp.	S
Azoxyanisole	117–134°C	4
4,4'-di-n-hexyloxyazoxy-benzene	70–125°C smectic at T ≈ 80°C	S di
p,p'-azoxyphenetole	138–168°C	m su
p-hexyloxycinnamic acid	152–179°C	F
$CH_3-O-C_6H_4-N=N-C_6H_4-O-CO-CH=CH-$ $-C_6H_4-R$ R = OCH_3 R = $-O-CO-O(C_2H_4O)_2C_2H_5$	 75–230°C 175°–340°C	 F ch
Arylene p-alkoxybenzoates	80–200°C	B
Cholesteryl myristate	70–82°C	X

[a]Additional examples of application are given in the extensive review by Kelker and von Schivizhoffen [1

chemical reactions. The first experiment of this type was performed long ago by Svedberg [137]. He studied the rate of thermal decomposition of picric acid in azoxyanisole. Orientation of the nematic phase by a magnetic field decreased the rate of decomposition. Moreover a sudden increase in the rate of picric acid decomposition was observed at the nematic-to-isotropic transition. Bacon and Brown [138] report an increase of the reaction rate of the Claisen rearrangement in a nematic state. Paleos and Labes [139] and Baturin et al. [140] studied polymerization reactions in nematic phases. The first authors did not find a significant effect of the nematic order (or of a magnetic field) on the rate of polymerization. The second group of authors, however, report a considerable increase in the rate of polymerization of p-methacryloxybenzoic acid with styrene when the reactants are ordered in a nematic phase [140]. More recently Barnett and Sohn [141] studied the pyrolysis of xanthate (a unimolecular reaction in which olefines are produced) in nematic solvents. These authors report an increase of the olefine production in the nematic state.

Chemical reactions depend strongly on the viscosity of the solvent. A comparison of the reaction rates in different mesomorphic phases of a solvent is therefore rather difficult. It is well possible that the reported changes in the reaction rates on going from an isotropic to a nematic solvent are due to viscosity changes. More experimental work is necessary before any conclusions concerning the influence of the partial orientation on the reaction rates can be drawn.

6.2. Possible Application in Analytical Chemistry

A well known property of cholesteric phases is the extreme sensitivity of the pitch (or the reflection color) against weak external perturbations such as temperature, mechanical stress or electric fields. The pitch P (and therefore the reflection color λ_{max} = nP) may also be disturbed considerably by small amounts of non-chiralic solute molecules. Fergasón [142] reported for the first time that the addition of traces of organic vapors (such as chloroform, methylene chloride, petroleum ether) may lead to observable color changes in cholesteric layers. Color changes are observed both if the liquid crystal merely serves as solvent or if the vapor reacts with the solvent molecules. Since small molecules diffuse rapidly into the liquid crystal, the vapors produce an immediate color response. The amount and the direction of the pitch (or color) change is a sensitive function of the cholesteric solvent. Consequently, any desired color change may be produced by using different mixtures of cholesteryl derivatives.

Another typical example for solute-induced pitch shifts is the color change produced in (cholesteric) mixtures of cholesteryl derivatives upon addition of *cis*- or *trans*-isomers of azobenzenes or stilbenes. In a cholesteric mixture of cholesteryl chloride and cholesteryl nonanoate (70 mole percent) cis azobenzene produces a red shift, while the trans-isomer induces a blue shift [143]. For cholesteric mixtures well above the cholesteric to smectic transition the shift $\Delta\lambda_{max}$ in the reflection color (λ_{max}) is proportional to the solute concentration [143, 144]:

$$\lambda_{max}^t = \lambda_{max}^o + q_t c_t; \lambda_{max}^c = \lambda_{max}^o + q_c c_c \tag{44}$$

In this equation c_t and c_c are the concentrations of trans- and cis-azobenzene, respectively and λ^o_{max} is the wavelength of maximum reflectivity of the pure solvent. The proportionality coefficients q_t and q_c are sensitive functions of the solvent composition. According to the above equations one expects that the color change induced by a mixture of cis- and trans-azobenzene is

$$\Delta\lambda_{max} = q_t c_t + q_c c_c$$

Such a behavior has been verified experimentally [143]. The above equation involves that the color change induced by a mixture of isomers is additively composed of the changes produced by the individual components. Accordingly, the composition of an unknown mixture of azobenzene isomers can be determined from the observed color change, provided the constants q_t and q_c (that are characteristic for the solvent) have been determined in a separate experiment. This example indicates a possible application of cholesteric phases for quantitative analytical work.

It is well known that transitions between cis- and trans-azobenzene can be induced photochemically [145] by excitation with appropriate wavelength (cf Fig. 20). Accordingly, colored pictures may be produced in a reversible way on cholesteric layers [143]. In the same way as azobenzene, the derivatives of azoxybenzene undergo *cis-trans* (or *trans-cis*) isomerization upon excitation and induce different color changes in cholesteric phases.

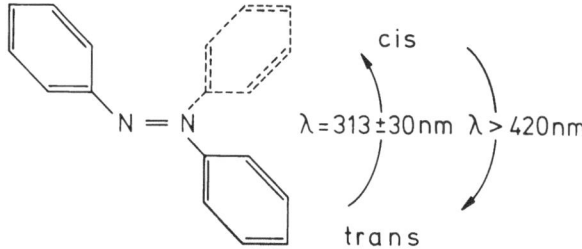

Figure 20. *Cis-trans* isomerization of azobenzene. λ is the excitation wavelength.

References

1. Kelker, H., Scheurle, B.: Angew. Chem. (Internat. Edit.) 8, 884 (1969).
2. Spiesecke, H., Jourdan, J. B.: Angew. Chem. (Internat. Edit.) 6, 450 (1967).
3. Pohl, L., Steinsträßer, R.: Z. Naturforsch. 26b, 26 (1971).
4. Steinsträßer, R.: Angew. Chem. 84, 636 (1972).
5. Englert, G., Saupe, A.: Mol. Cryst. 1, 503 (1966).
6. Meiboom, S., Luz, Z.: Fourth International Liquid Crystal Conference, Kent, Ohio (1972).
7. Sackmann, E., Möhwald, H.: J. Chem. Phys. 58, 5407 (1973).
8. Sackmann, E., Meiboom, S., Snyder, L. C., Meixner, A. E., Dietz, R. E.: J. Am. Chem. Soc. 90, 3567 (1968a).
9. Robinson, C.: Mol. Cryst. 1, 467 (1966).
10. Panar, W., Phillips, W. D.: J. Am. Chem. Soc. 90, 3880 (1968).
11. Samulski, E., Tobolsky, A. V.: Mol. Cryst. and Liq. Cryst. 7, 433 (1969).
12. Lawson, K. D., Flautt, T. J.: J. Am. Chem. Soc. 89, 5489 (1967).

13. Black, P. J., Lawson, K. D., Flautt, T. J.: J. Chem. Phys. **50**, 542 (1969).
14. Mandell, L., Fontell, K., Ekwall, P.: Adv. Chem. Ser. **63**, 89 (1967).
15. Seelig, J.: J. Am. Chem. Soc. **92**, 3881 (1970).
16. Cladis, P. E., Rault, J., Burger, J.-P.: Mol. Cryst. and Liq. Cryst. **13**, 1 (1971).
17. Brière, G., Gaspard, F.: Chem. Phys. Letters **1**, 706 (1968).
18. Borrett, S., Brière, G.: J. Chim. Phys. **9**, 970 (1965).
19. Berremen, D.: Phys. Rev. Letters **28**, 1683 (1972).
20. Caesar, G. P., Gray, H. B.: J. Am. Chem. Soc. **91**, 191 (1969).
21. Maier, W., Englert, G.: Z. Elektrochemie **62**, 1020 (1958).
22. Kelker, H., Hatzand, R., Wirzing, G.: Z. Analyt. Chem. **267**, 161 (1973).
23. Haas, W., Adams, J., Flannery, J. B.: Phys. Rev. Letters, **25**, 1326 (1970).
24. Mainusch, K.-J., Müller, U., Pollmann, P., Stegemeyer, H.: Z. Naturforsch. **27A**, 1677 (1972).
25. Helfrich, W., Oh, Ch. S.: Mol. Cryst. and Liq. Cryst. **14**, 289 (1971).
26. Sackmann, E., Meiboom, S., Snyder, L. C.: J. Am. Chem. Soc. **89**, 5981 (1967).
27. Sackmann, E.: Chem. Phys. Letters **3**, 253 (1969).
28. Bässler, H.: Festkörperprobleme XI, (O. Madelung Ed.), Vieweg/Pergamon: 1971.
29. Wysocki, J. J., Adams, J., Haas, W.: Phys. Rev. Letters **20**, 1024 (1968).
30. Meyer, R. B.: Appl. Phys. Letters **9**, 281 (1968).
31. De Gennes, P. G.: Solid State Commun. **6**, 163 (1968).
32. Bässler, H., Labes, M. M.: J. Chem. Phys. **51**, 1846 (1969).
33. Diehl, P., Khetrapal, C. L., Kellerhals, H. P., Lienhard, U., Niederberger, W.: J. Magn. Res. **1**, 527 (1969).
34. Buckingham, A. D., Pople, J. A.: Trans. Farad. Soc. **59**, 2421 (1963).
35. Ducros, P.: Bull. Soc. Franc. Mineral. Crist. **83**, 85 (1960).
36. Andrew, E. R., Wynn, V. T.: Proc. Roy. Soc., **291A**, 257 (1966).
37. Saupe, A., Englert, G.: Phys. Rev. Letters **11**, 402 (1963).
38. Englert, G., Saupe, A.: Z. Naturforsch. **19a**, 172 (1964).
39. Saupe, A.: Angew. Chem. **80**, 99 (1969).
40. Meiboom, S., Snyder, L. C.: Science **162**, 1337 (1968).
41. Luckhurst, G. R.; Quart. Rev. Chem. Soc. **22**, 179 (1968).
42. Hilbers, C. W., Mac Lean, C.: In N.M.R. Basic Principles and Progress Vol. 7 (Diehl, P., Fluck E., Kosfeld, R., Eds) (Springer-Verlag Berlin) 1972.
43. Diehl, P., Khetrapal, C. L.: N.M.R. Basic Priniciples and Progress Vol. 1 (Diehl, P., Fluck, E., Klosfeld, R., Eds.) (Berlin Springer-Verlag) 1970.
44. Meiboom, S., Snyder, L. C.: Acc. Chem. Res. **4**, 81 (1971).
45. Snyder, L. C.: J. Chem. Phys. **43**, 4041 (1965).
46. Buckingham, A. D., Love, I.: J. Magn. Resonance **2**, 338 (1970).
47. Krugh, T. R., Bernhein, R. A.: J. Am. Chem. Soc. **91**, 2385 (1969).
48. Herschbach, D. R., Laurie, V. W.: J. Chem. Phys. **37**, 1668 (1962).
49. Saupe, A., Englert, G., Povh, A.: Adv. Chem. Ser. **63**, 51 (1967).
50. Meiboom, S., Snyder, L. C.: J. Am. Chem. Soc. **89**, 1038, 5981, (1967).
51. Bastiansen, O., Fritsch, F. N., Hedgerg, K.: Acta Cryst. **17**, 538 (1964).
52. Sackmann, E.: J. Chem. Phys. **51**, 2984 (1969).
53. Gazzard, J. J., Sheppard, N.: Mol. Phys. **21**, 169 (1971).
54. Cocivera, M.: J. Chem. Phys., **47**, 3061 (1967).
55. Diehl, P., Henrichs, P. M., Niederberger, W.: Mol. Phys. **20**, 139 (1971).
56. Sackmann, E., Meiboom, S., Snyder, L. C.: J. Am. Chem. Soc., **90**, 2183 (1968).
57. Luckhurst, G. R.: In Liquid Crystals and Plastic Crystals (Gray, G. W., Winsor, P. A. Eds) (Ellis Horwood Limited Chichester) (1974).
58. Luckhurst, G. R.: Fourth International Liquid Crystal Conference, Kent, Ohio, USA (1972).
59. Krebs, P., Sackmann, E.: Mol. Phys. **23**, 437 (1972).
60. Krebs, P.: Doctoral Thesis, Stuttgart (1973).
61. Hubbell, W. L., McConnell, H. M.: J. Am. Chem. Soc. **93**, 314 (1971).
62. Jost, P., Waggoner, A. S., Griffith, O. H.: In: "Structure and Function of Biological Membranes" (Rothfield, L. J., Ed.) New York: Academic Press pp. 84–144 1971.
63. Sackmann, E., Träuble, H.: J. Am. Chem. Soc. **94**, 4482 (1972).

64. Smith, I. C. P.: "The Spin Label Method" in Biological Application of Electron Spin Resonance. (Swartz, H. M., Bolton, J. R., Brog, D. C., Eds.) New York: Wiley Interscience, pp. 483–539 1972.
65. Falle, H. R., Luckhurst, G. R.: J. Magn. Resonance 3, 161 (1970).
66. Falle, H. R., Luckhurst, G. R., Lemaire, H., Morat, C., Rassat, A., Rey, P.: Mol. Phys. 11, 49 (1966).
67. Agranat, I., Rabinovitz, M., Falle, H. R., Luckhurst, G. R., Ockwell, J. N.: J. Chem. Soc. (1970), 294.
68. Lemaire, H.: J. Chim. Phys. 64, 559 (1967).
69. Krebs, P., Sackmann, E., Voss, J.: Chem. Phys. Letters 8, 417 (1972).
70. Glarum, S. H., Marshall, J. H.: J. Chem. Phys. 44, 2884 (1966).
71. Glarum, S. H., Marshall, J. H.: J. Chem. Phys., 46, 55 (1967).
72. Möbius, K., Haustein, H., Plato, M.: Z. Naturforsch. 23a, 1626 (1968).
73. McConnell, H. M., Strathdee, J.: Mol. Phys. 2, 129 (1959).
74. McLachlan, A. D.: Mol. Phys. 3, 233 (1960).
75. Hall, G. G., Hardisson, A.: Proc. Roy. Soc. London, 278A, 129 (1964).
76. Stone, A. J.: Mol. Phys. 6, 509 (1963); 7, 311 (1964).
77. Snyder, L. C., Amos, T.: J. Chem. Phys. 41, 1773 (1964); 42, 3670 (1965).
78. Haustein, H., Möbius, K., Dinse, K. P.: Z. Naturforsch. 24a, 1768 (1969).
79. Dinse, K. P., Biehl, R., Möbius, K., Haustein, H.: Chem. Phys. Letters 12, 399 (1971).
80. Hutchison, C. A., Mangum, B. W.: J. Chem. Phys. 29, 952 (1958).
81. Kottis, P., Lefebvre, R.: J. Chem. Phys. 41, 3660 (1964).
82. van der Waals, J. H., de Groot, M. S.: Mol. Phys. 3, 190 (1960).
83. Kottis, P.: Ann. Physique 4, 459 (1960).
84. Heppke, G., Schneider, F.: Ber. Bunseng. Physik. Chem. 75, 61 (1971).
85. McClure, D. S.: J. Chem. Phys. 22, 1668 (1954).
86. Labhart, H.: Adv. Chem. Phys. 13, 179 (1967).
87. Liptay, W.: Angew. Chem. 81, 195 (1969).
88. Weber, G.: Biochem. J. 75, 335 (1960).
89. Albrecht, A. C.: J. Mol. Spectrosc., 6, 84 (1961).
90. Zimmermann, H., Joop, N.: Z. Elektrochem. 66, 342 (1962).
91. Zimmermann, H., Joop, N.: Z. Elektrochem. 65, 66 (1962).
92. Dörr, F.: Angew. Chem. 78, 457 (1966).
93. Dörr, F.: In "Creation and Detection of the excited state" (Lamola, A. A., Ed.) Vol. I New York. M. Dekker, pp. 53–122 (1971).
94. Jablonski, A.: Acta Phys. Polon. 3, 421 (1934).
95. Thulstrup, E. W., Michl, J., Eggers, J. H.: J. Phys. Chem. 74, 3868 (1970).
96. Thulstrup, E. W., Eggers, J. H.: Chem. Phys. Letters 1, 690 (1968).
97. Dörr, F., Gropper, H.: Ber. Bunseng. Physik. Chem. 67, 193 (1963).
98. Sackmann, E., Voss, J.: Chem. Phys. Letters 14, 528 (1972).
99. Beens, H., Möhwald, H., Rehm, D., Sackmann, E., Weller, A.: Chem. Phys. Letters 8, 341 (1971).
100. Baur, G., Stieb, A., Meier, G.: Mol. Cryst. and Liq. Cryst. in press (1973).
101. Chalelain, P.: Acta Cryst. 4, 435 (1951).
102. Hochstrasser, R. M.: "Molecular Aspects of Symmetry" New York: W. A. Benjamin 1966.
103. Birks, J. B.: "Photophysics of Aromata Molecules" London: Wiley-Interscience 1970.
104. Saupe, A., Maier, W.: Z. Naturforsch. 16a, 816 (1961).
105. Saeva, F. D., Wysocki, J. J.: J. Am. Chem. Soc. 93, 5928 (1971).
106. De Vries, H.: Acta Cryst. 4, 219 (1966).
107. Mauguin, C.: Bull. Soc. Franc. Mineral 34, 71 (1911).
108. Stegemeyer, H., Mainusch, K. J.: Chem. Phys. Letters. 16, 38 (1972).
109. Fergason, J. L.: Mol. Cryst. and Liq. Cryst. 1, 293 (1966).
110. Mainusch, K.-J., Pollmann, P., Stegemeyer, H.: Ber. Bunseng. Physik. Chem. 77, (1973).
111. Sackmann, E., Rehm, D.: Chem. Phys. Letters 4, 537 (1970).
112. Caesar, G. P., Levenson, R. A., Gray, H. B.: J. Am. Chem. Soc., 91, 772 (1969).

113. Flitcroft, N., Huggins, D. K., Kaesz, H. D.: Inorg. Chem. **3**, 1123 (1964).
114. Maier, W., Saupe, A.: Z. Naturforsch. **14a**, 882 (1959).
115. Maier, W., Saupe, A.: Z. Naturforsch. **15a**, 287 (1960).
116. Saupe, A.: Mol. Crystals **1**, 527 (1966).
117. Krieger, T. J., James, H. M.: J. Chem. Phys. **22**, 796 (1954).
118. McMillan, W. L. Phys. Rev. **A4**, 1238 (1971).
119. Chen, D. H., James, P. G., Luckhurst, G. R.: Mol. Cryst. and Liq. Cryst. **8**, 71 (1969).
120. Humphries, R. L., James, P. G., Luckhurst, G. R.: J. Chem. Soc. (Farad. Trans. II) **68**, 737 (1972).
121. Saupe, A., Nehring: Mol. Cryst. Liq. Cryst. **8**, 403 (1969).
122. Lefebvre, C.: In: "Physical methods of organic chemistry (A. Weissberger Ed.) New York: Interscience (1960) Ch. 36.
123. Sackmann, E., Krebs, P., Rega, H. U., Voss, J., Möhwald, H.: Mol. Cryst. and Liq. Cryst. **24**, 283 (1973).
124. Robertson, J. C., Yim, C. T., Gilson, D. F. R.: Canad. J. Chem. **49**, 2346 (1969).
125. Wulf, A.: J. Chem. Phys. **55**, 4512 (1971).
126. Chandrasekhar, S., Madhusudana, M. V.: Acta Cryst. **A27**, 303 (1971).
127. Dewar, M. J. S., Schröder, J. P.: J. Am. Chem. Soc. **86**, 5235 (1964).
128. Chow, L. C., Martire, D. E.: J. Phys. Chem. 75, 2005, (1971).
129. Humphries, R. L., Luckhurst, G. R.: Faraday Symposium No. 5, Liquid Crystals (1971).
130. Zielinski, W. L., Freeman, D. H., Martire, D. E., Chow, L. C.: Anal. Chem. **42**, 176 (1970).
131. Kelker, H.: Ber. Bunsenges. 67, 698 (1963).
132. Kelker, H.: Z. Analyt. Chem. 198, 254 (1963).
133. Kelker, H., von Schivizhoffen, E.: Adv. Chromatography, (J. C. Giddings and R. A. Keller Eds.) 6, 247 (1968).
134. Kelker, H., Verhelst, A.: J. Chromatogr. Sci. **7**, 79 (1969).
135. Kelker, H., Scheurle, B., Sabel, J., Jainz, J., Winterscheidt, H.: Mol. Cryst. and Liq. Cryst. **12**, 113 (1971).
136. Martire, D. E., Blasco, P. A., Carone, P. F., Chow, L. C., Vicini, H.: J. Phys. Chem. **72**, 3489 (1968).
137. Svedberg, T.: Kolloid Z. 18, 54, 101 (1916).
138. Bacon, W. E., Brown, G. H.: Mol. Cryst. and Liq. Cryst. **6**, 155 (1969).
139. Paleos, C. M., Labes, M. M.: Mol. Cryst. and Liq. Cryst. **11**, 385 (1970).
140. Baturin, A. A., Amerik, Y. B., Krentsel, B. A., Mol. Cryst. and Liq. Cryst. **16**, 117 (1972).
141. Barnett, W. E., Sohn, W. H.: Chem. Communications **1002** (1971).
142. Fergason, J. L.: Scientific American, **211**, 76 (1964).
143. Sackmann, E.: J. Am. Chem. Soc. **93**, 7088 (1971).
144. Voss, J., Sackmann, E.: Z. Naturforsch. **28A**, 1496 (1973).
145. Zimmerman, G., Chow, L., Paik, U.: J. Am. Chem. Soc. **80**, 3528 (1958).

Medical and Technical Applications of Liquid Crystals

J. G. Grabmaier

Forschungslaboratorien der Siemens AG,

D-8000 München, Federal Republic of Germany

1. Application of Cholesteric Liquid Crystals

1.1. Introduction

Certain cholesteric liquid crystals exhibit color changes over the entire color scale from red to violet when a variable of state such as the temperature of the liquid crystal is changed over a particular range. The process is normally reversible, and the colors occur in the reverse sequence with falling temperature to that with rising temperature. A simple method is therefore presented of measuring surface temperatures, of utilizing color differences to indicate temperature variations, and of following dynamic processes such as thermal diffusion. Similar color shifts are also obtainable by changing electric or magnetic fields, by changing the pressure and by changing the concentration of gases or vapors.

As the intermolecular dispersion forces that maintain the molecules in the cholesteric structure are relatively small, a low additional amount of energy suffices to change the helix pitch. The pitch depends particularly on temperature. In some cases the dependence is so pronounced that typically a temperature rise of less than $3°C$ is adequate to change the color of reflected light from red over the entire visible spectrum to blue. The use of cholesteric liquid crystals in thermographic applications is based on this property (see also first part of this book).

1.2. Medical Applications

Cholesteric liquid crystals can be used in thermal mapping of human skin both for basic studies of the circulatory system and for diagnosis of circulatory system diseases or for detection of tumors. The basic idea is that areas of the body in which circulation is poor will have lower temperatures than areas with good circulation [1].

1.2.1. Diagnosis of Vascular Diseases

The surface temperature of the human skin under which vessels are located is about 0.1°C to 0.3°C higher than that of the surrounding area. This is particularly true of the limbs, where the vessels are relatively close to the surface. Sensitive thermographic methods can then be employed to follow the course of vessels and localize the points where the blood supply to the respective limb is impaired as a result of calcification or a clot. By using a cholesteric liquid crystal with an indicating range of 0.4°C or less, the color of which is blue at normal body temperature, the diseased points will show up red because of the lower temperature [2].

1.2.2. Cancer Diagnosis

The rapid reproduction of cancer cells is associated with increased metabolism and hence higher temperature than exhibited by cells with a normal growth function. If those parts of the body where suspected cancer is located are coated with a cholesteric liquid crystal, the existence of cancer is indicated by the local color change as a result of the higher temperature [1, 2]. The liquid crystal thermograph is therefore an ideal means for early detection of cancer and diagnosis of cancerous tumors. The centre of the growth can, however, only be accurately localized by this thermographic process when the tumor is close to the skin surface, for example skin or breast cancer. The thermographic findings can in fact be used to differentiate between benign and malignant tumors [3].

The difference in temperature between the tumor and the surrounding skin is usually 0.7°C to 2.5°C, sometimes higher, depending on the type of cancer. The temperature gradient can be increased by creating thermal stress, for instance blowing the area with cold or hot air. The subjectively disagreeable blackening and spraying of the patient is compensated by the relatively rapid completion of the test and the low cost.

1.2.3. Localizing the Placenta

When complications in the later stages of a pregnancy necessitate an operation, for example a Caesarian, the position of the placenta must be accurately determined immediately prior to the operation. Here the fact can be exploited that the temperature of the skin region above the placenta is about 2°C higher than that of the surrounding area. Foil pads containing a cholesteric liquid crystal that changes color between 31°C and 35°C are best used as indicators [4].

1.2.4. Pharmacological Tests

Certain pharmacological tests involving introduction of chemical substances into the blood stream produce allergic skin reactions due to liberation of histamine. These reactions are associated with increased development of heat. The reaction rate of such substances can be determined by checking the temperature with cholesteric temperature indicators [5]. The affect of medicaments to combat allergic reactions can also be followed.

1.2.5. Skin Grafting

In plastic and restorative surgery there are often problems in covering larger areas of defective skin. If the area is very large, so-called stalked flaps are used to cover the defect. This standard practice in skin surgery involves transplanting large pieces of skin from one part of the body to another, with the pieces of skin kept connected temporarily during the transplantation phase to their original locations by a stalk.

An important aspect in plastic surgery is to improve the blood supply to the stalked flaps before and after grafting. A further problem is to determine at which point in time the connecting skin supplying nutrition should be severed. It has become evident that this point is reached by men when the skin flap temperature has reached $29°C$ [6]. The flap temperature is measured with cholesteric liquid crystals that change color with increasing temperature over the entire spectrum from red to blue and react to small temperature changes.

1.3. Technical Applications

Because of their unique color properties, cholesteric liquid crystals can be employed to indicate temperature field patterns and for color picture screens [7, 8]. However, as cholesteric material has a minimum reaction time which is in the order of 0.1 s, it is only suitable for displaying color information when the information changes slowly. It is unsuitable for reproducing fast changing, moving pictures.

1.3.1. Direct Temperature Diagrams

There are two possibilities of obtaining direct temperature diagrams of objects: either the infrared radiation from the object is focused by a suitable optical system onto an infrared-absorbant layer and this layer is in good thermal contact with a cholesteric liquid crystal film, or the object is brought into direct thermal contact with a cholesteric liquid crystal film. The cholesteric film then converts the temperature diagram of the object into visible form. This can either be directly viewed by illuminating with a suitable light source or automatically evaluated.

1.3.1.1. Infrared Display Unit

The above concept can be utilized to build an infrared display unit [9] (Fig. 1) consisting basically of a temperature-controlled chamber that can be evacuated and which has windows at opposite ends. One window is transparent to infrared wavelengths and is made, for example, of sodium chloride. The other window is transparent to wavelengths between 400 nm and 750 nm, i.e. the visible spectrum, and is made of glass. Between these two windows and parallel to them the actual image converter is located: a polyester foil about 5 μm thick is coated on the incident radiation side with a 1 μm to 3 μm layer capable of quantitatively absorbing infrared radiation, for example finely distributed gold or nickel. The other side is

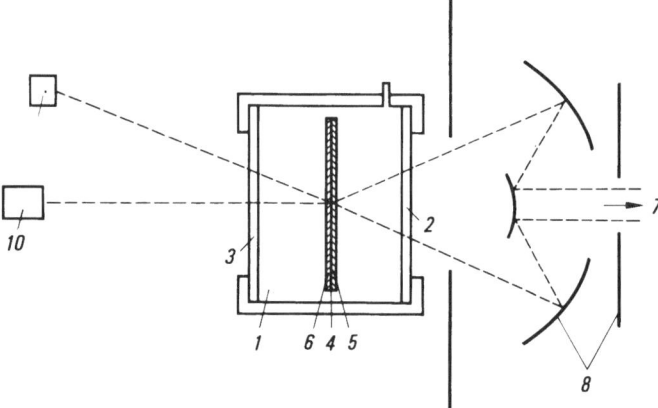

Figure 1. Scheme of an infrared display unit based on cholesteric liquid crystals (after reference [9]).
 1 Chamber
 2 NaCl-window (IR-translucent)
 3 Glass-window (translucent for visible light)
 4 Supporting
 5 Heat absorbing layer (Au, Ni)
 6 Cholesteric layer
 7 Object
 8 IR-optic (mirrors and diaphragm)
 9 Light source
10 Observer

coated with a 10 μm layer of cholesteric material. Prior to irradiation the heat-sensitive layer should have a uniform surface temperature a few tenths of a degree below the temperature at which the cholesteric film turns red. This controlled temperature can be accomplished by evacuating the complete chamber. If by these conditions the infrared radiation from the object is focused by a suitable infrared optic onto an infrared absorbant layer, which is in good thermal contact with a cholesteric liquid crystal film the film then converts the temperature diagram of the subject into visible form if illuminated with a light source.

1.3.1.2. Wind Tunnel Experiments

The traditional woollen thread method employed when investigating the aerodynamic properties of flying objects is now being replaced by thermographic measuring techniques [10]. Use is made of the condition that turbulences produce more local frictional heat than laminar flow. The temperature difference between regions of turbulent and laminar flow in a wind tunnel test at Mach 1.6 to 3 is less than 3°C [11]. Quick results can be obtained by coating the model with a cholesteric liquid crystal film to act as a temperature indicator. As the colors are reversible, it is possible by changing e.g. the angle of the model to the air current to simulate flight conditions and measure the flying characteristics in every flight attitude.

The test results are analyzed photographically. Whereas a single picture shows exactly the regions of higher and lower temperature because of the differing colors, evaluating a series of photographs yields more detail; that point of a turbulent region where heating first occurs can be detected. This generally makes it possible to localize the cause of a turbulence and also to clear it.

1.3.1.3. Intensity Distribution of Wave Fields

Microwave Fields

With the help of cholesteric liquid crystals, intensity distributions of microwave fields can be rendered visible [12]. A suitable detector (Fig. 2) consists of a thin resistive metal film with a surface resistance of 200 $\Omega/\square$, to 400 $\Omega/\square$ coated with a 10 μm to 20 μm layer of a cholesteric liquid crystal. A polyester film deposited on both sides protects the device. When this foil is introduced into a microwave field of sufficient power density, absorption in the metal film produces a temperature profile proportional to the field intensity distribution. The liquid crystal temperature indicator then makes this profile visible.

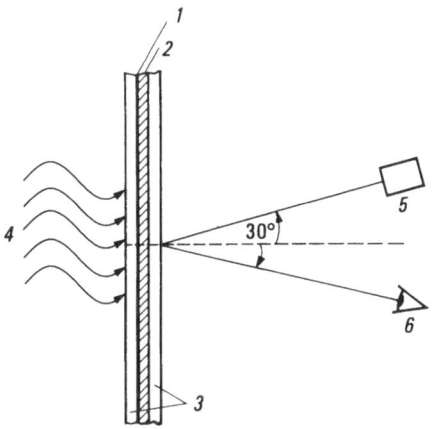

Figure 2. Scheme of a microwave field-detector based on cholesteric liquid crystals (after reference [12]).

1 Resistance layer (200 $\Omega/\square$–400 $\Omega/\square$)	4 RF field
2 Cholesteric layer (10μ m–20 μm)	5 Light source
3 Polyester foil (10 μm–30 μm)	6 Observer

Sound Fields

Cholesteric liquid crystals can also be used to display sound fields, especially ultrasonic fields [13, 14]. The ultrasonic source is immersed in water and directed at a black foil on the water surface. The upper side of the foil is coated with a cholesteric liquid crystal film. The acoustic irradiation of the indicator foil generates colored images of the ultrasonic source and also various interference patterns. The reason for colors appearing can be attributed to conversion to heat of the ultrasonic energy absorbed by the foil.

Similarly the sound fields produced by an electro-acoustically excited membrane can be demonstrated by cholesteric liquid crystals. In this case, however, both thermal and mechanical reactions cause the color changes.

It is important for these applications that the ambient temperature is closely stabilized and preferably slightly lower than the temperature at which the cholesteric liquid crystal is red. This field detector is ideal for studying the short-range field of microwave antennas where the power density is high enough.

1.3.1.4. Fault Location in Electronic Devices

Based on the generation of a temperature pattern by dissipation of power, liquid crystals can be employed to detect faults in electronic devices. The liquid crystalline material is painted on the surface of the device and the device is then switched on. The defective areas are clearly indicated by the change in color of the liquid crystal film at the points of excessive thermal stress. This technique is particularly useful for locating faults in solid state devices, especially integrated circuits [15].

1.3.2. Indirect Temperature Patterns

1.3.2.1. Thermally Activated Information Displays by Tunnel Currents

If information is converted to x,y coordinates and transmitted by scanning to a picture screen of cholesteric liquid crystals, localized heating can be obtained on the screen corresponding to the electronic information. When the points of localized heating have different heat content, a colored picture is formed.

When two metals across which a potential difference exists are separated by less than 10 nm, a tunnel current flows through them. This tunnel current does not increase linearly but exponentially with the applied voltage. An XY electrode raster (Fig. 3) consisting, for example, of aluminum strips 400 μm wide and 10 nm thick separated by a 5 nm thick Al_2O_3 layer (Fig. 3a) in thermal contact with a cholesteric liquid crystal layer forms a matrix-type display screen [16]. If a potential of 2 V to 3 V is maintained between the two electrodes at one raster point, slight variations in potential as occurring when information is written in, cause large changes in current (Fig. 3b) and hence in temperature at this raster point. Concerning reproduceable results the XY matrix display is arranged on an Al-block.

1.3.2.2. Thermally Activated Information Displays by Photocurrent

An image converter or image intensifier with a liquid crystal layer consists of a photoconductive layer and a cholesteric liquid crystal layer embedded between two electrodes. If the two electrodes are subjected to an electric field, the light from a primary image falling on the photoconductor causes a change in conductivity at that point. Depending on the local brightness of the primary image a thermal pattern forms on the photoconductor as a result of the Joule effect. With good thermal contact between the photoconductive and liquid crystal layer, this thermal pattern can be utilized to display a secondary image on a cholesteric liquid crystal screen [16]. This principle is ideal for enlarging a primary image of a neutral television screen.

1.3.3. General Aspects in Using Cholesterics for Color Mapping

The specific color changes for a given compound depend, to a complex degree, on its molecular structure. Some empirical rules relating the structure to color sensitivity have been formulated. For example, derivatives of cholesterol with long

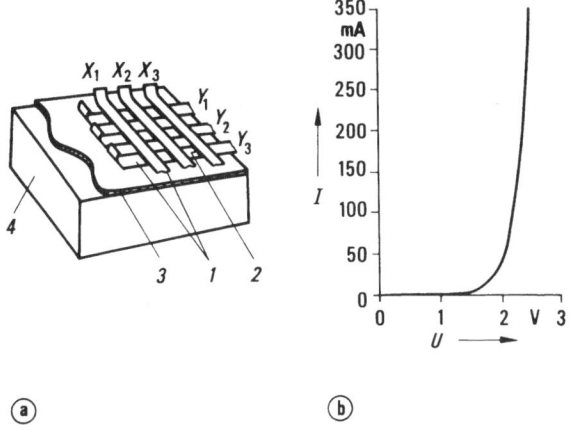

Figure 3. Thermally-activated X-Y-matrix display (after reference [16]).
(a) Schematic setup
(b) Nonlinear voltage-current characteristic I = f(U) of an Al-Al$_2$O$_3$-Al-element
1   Al-strips   3   Isolating layer
2   Al$_2$O$_3$-layer   4   Al-block

side chains exhibit a high sensitivity of color to temperature changes, whereas those
with short chains show low sensitivity with temperature. Intermediate temperature
sensitivity can be achieved with mixtures of short and long chain compounds [7].

In practice, cholesteric liquid crystals are of interest with a particular color at a
certain temperature or with a particular change in color over a certain temperature
range. The ternary mixture cholesteryl pelargonate – cholesteryl propionate –
cholesteryl oleyl carbonate, for example, produces a temperature indicator which
changes color from red at low temperature to blue at high temperature in the range
0 to 25°C depending on the mixture ratio. The temperature interval for the change
in color varies between 10°C and about 6°C. Mixtures of special cholesteryl esters
can be made to change color from red to blue within 0.5°C. Such indicators should
achieve a temperature resolution of at least 0.05°C [5]. To obtain good results,
working in an air-conditioned room is a prerequisite.

To make the color change clearly visible, the surface to which the liquid crystal
film is applied must be black or first dyed black. If the liquid crystal is applied to a
pale background, light is reflected not only from the liquid crystal but also from
the test object. Generally the surface is coated with a black varnish but sometimes
black foil is stuck on the surface. In all events the surface must be resistant to
liquid crystals or made so by coating with a film not dissolved by liquid crystals such
as polyvinyl alcohol [17].

The cholesteric temperature indicator is usually sprayed or painted on in a 10%
solution. The indicator solvent must have completely evaporated before reproducible
results can be expected.

A cholesteric liquid crystal layer normally disintegrates into smaller discrete
homogeneously aligned regions. In this case only less brilliant colors are obtained.

The brilliance of the colors can be appreciably increased by homogeneous alignment, especially homogeneous planar alignment of the liquid crystal layer. This can be accomplished by stroking over the liquid crystal layer once or twice in the same direction with a wide flat spatula or brush after the solvent has evaporated [18].

When utilizing the color changes of cholesteric liquid crystals an important factor that must be considered is that the wavelength of maximum scattering is a function not only of the angle of incidence of light but also of the viewing angle of the scattered light. It is therefore always important to perform measurements at the same viewing angle. The area to be investigated is illuminated with white light approximately normal to its surface and viewed at angles of 20° to 30° to the normal [18]. The color phenomena of cholesteric liquid crystals are best photographed by illuminating the object with a strong light source about 1 m from the surface and normal to it. The camera is located at the same distance and at about the same angle. The surface reflections of the cholesteric layer are screened by a polarization filter [17, 18]. Changing the temperature or the ambient conditions of the system will cause the wavelength for maximum scattering to shift from ultraviolet to infrared or vice versa. The reason for this shift is a change in pitch of the characteristic helix of the structure.

Cholesteric liquid crystals can be used to measure the absolute temperature of a process provided proper calibration can be performed. Several calibration techniques have already been developed. Perhaps the most obvious is to keep the system temperature constant and measure the reflectance at several points with a spectrophotometer.

Successful utilization of the temperature-dependent color changes of cholesteric liquid crystals assumes several prerequisites are met with [7, 17]:

1. The heat capacity of the object to be studied should be larger than that of the cholesteric film. As the heat capacity of most cholesteric liquid crystals is in the order of 1.5 J/cm^3 and sheets of about 20 μm are usually employed, the heat capacity is about 3×10^{-3} J/cm^2. Unless the test specimen is a very thin film, this value is sufficiently low.

2. The specimen to be examined must be larger than the resolution of the liquid crystal layer, which is about 20 μm.

3. The rate of temperature change of the specimen to be examined must not be less than 0.1 s so that the liquid crystal can follow it.

4. The temperature variation of the specimen during the test must lie within the range covered by the liquid crystals available, which is e.g. 0°C to 75°C.

Liquid crystal methods offer numerous advantages over other techniques. Most obvious is the low cost. No expensive and elaborate equipment is needed; all that is required are a few bottles of material, a means of applying it to the specimen and a way of applying heat or extracting it from the specimen. Very often there is no other method of thermal mapping.

There are also some disadvantages and limitations. Perhaps the most apparent is the relatively low temperature at which liquid crystals can be used. In air the limit is about 190°C, in the absence of air 270°C. Moreover, the high temperature materials crystallize and tend to separate from the solution at room temperature; this makes them more difficult to apply. Another disadvantage is the short lifetime of liquid crystals due to chemical decomposition when exposed to air, particularly at elevated temperatures. At room temperature the limit is about three to four hours. However, methods of encapsulating the material have been developed to overcome these problems. In the National Cash Register process the material is encapsulated in spheres approximately 50 μm in diameter [19]

The encapsulated material can be used like the pure material in many applications. The two disadvantages — slight loss in resolution and increase in heat capacity per unit area because of the need to use thicker films — are not serious enough to prevent using the encapsulating material where a particular test procedure necessitates long-term stability.

The response time of cholesteric liquid crystals is finite and comparatively long because the visible color change occurs in two stages. An instantaneous change in the temperature of a cholesteric liquid crystal causes it to assume a new pitch. However, this requires that some of the material flows to a different configuration. The rate at which such flow occurs is limited by the material viscosity in the direction of flow.

The time constant of ordinary cholesteric materials is in the order of 0.1 s. For example, for cholesteryl nonanoate it is 0.1 s and for cholesteryl oleyl carbonate 0.2 s. These response times do not indicate a limitation if the materials are to be observed by eye: motion pictures at 64 frames/s have been successfully made and information determined from single frames.

1.3.4. Electrically Controllable Color Switching in Cholesterics

1.3.4.1. Electric Field-induced Imaging Process and Color Change in Cholesteric Liquid Crystal Films

In addition to depending on temperature, the molecular alignment of a cholesteric liquid crystal can be disturbed by electric fields resulting in a change of the pitch of the helical structure, which yields a profound effect on reflection [20]. If a voltage is applied to a thin liquid crystal film the magnitude of the voltage determines the spectral position of its scattering band, as illustrated in Fig. 4a. Since this scattering band is only some 10 nm in width, relatively pure reflected spectral colors are observed with white-light illumination. The particular color is determined by the magnitude of the voltage applied across the film. Fig. 4b illustrates the simplest form of how an increasing voltage changes the effective pitch of the planar texture, so that increasing the voltage will move the peak of the scattering band towards blue. If the sample is illuminated by monochromatic light and the voltage is varied so that the wavelength of maximum scattering passes through the wavelength of the source, the intensity of light scattered will increase to a maximum and then decrease.

A schematic cross section of a screen based on this principle is shown in Fig. 5. A thin layer of cholesteric liquid crystal is sandwiched between a bottom layer of polyester blackened on the opposite side and a top layer of transparent polyester. The black layer serves both as a conducting lower electrode (approx. 1 kΩ per square) and as an absorber for the transmitted light. The bottom polyester layer presents an orienting surface for the liquid crystal alignment. The top polyester

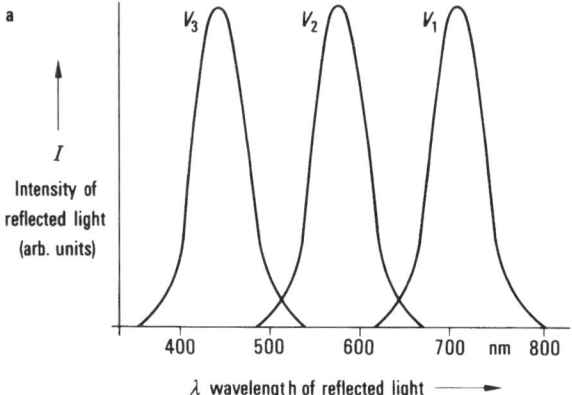

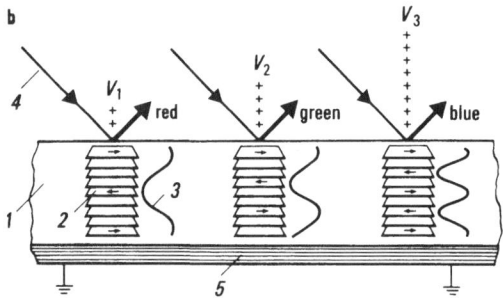

Figure 4. Effect of voltage on a cholesteric liquid crystal layer (after reference [20]).
(a) Spectral reflectance for voltages $V_3 > V_2 > V_1$, white illumination, constant viewing and illumination angle
(b) Relationship between voltage and effective pitch

1   Liquid crystal film   4   Incident white light
2   Schematic of helical structure   5   Conducting block layer
3   Pitch

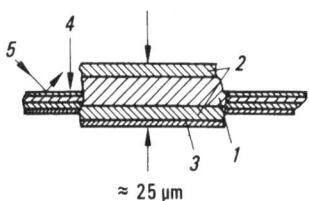

Figure 5. Schematic of polyester-liquid crystal sandwich (after reference [20]).
1   Cholesteric liquid crystal layer   4   Electron beam
2   Polyester   5   Incident light
3   Conducting black

layer allows an electric charge to be built up on the free surface of the polyester by a scanning electron beam with little lateral charge leakage. Alternatively it is coated with a transparent electrode so that a voltage can be applied to the electrodes. Because a top-and-bottom orienting "sandwich" structure enhances the reflected brightness, the top polyester layer also presents an orienting surface for the liquid crystal alignment. For operation the system is placed in a continuously pumped vacuum tube.

Because electrostatic fields can be created and controlled by a variety of methods, including electron beam scanning, the method described here may be suitable for color recording on picture screens. The main disadvantages of this screen are thermal and electrical hysteresis phenomena of the cholesteric liquid crystal layer.

1.3.4.2. Electrophotographic Imaging with Cholesteric Liquid Crystals

In imaging methods based on the deformation of an insulating viscous film, an electrostatic pattern becomes visible. With a cholesteric liquid crystal film, deformation of the film caused by an electrostatic pattern yields a colored picture. As the electrostatic pattern is controllable by incident light, electrophotographic imaging is obtained [21].

The knowledge of two processes is important to understand this imaging technique: the Eidophor process and the xerographic process. In the Eidophor process, a thin insulating oil film on a plate of insulating material with non-uniform electrostatic charge distribution is deformed as a result of an electron beam discharging the substrate. The deformation image is projected onto a screen. During deformation the oil film shifts to regions where the charge carrier concentrations are low.

In the xerographic process a plate of photoconducting amorphous selenium is charged to a uniform potential by a corona discharging device. The selenium plate is then discharged in an exposure step. After exposure an electrostatic pattern exists on the selenium surface corresponding to the image which is made visible (developed) with charged black toner powder attracted by the electrostatic surface charge.

In the electrophotographic imaging process with cholesteric liquid crystals, a standard selenium plate 50 μm thick on a metal substrate is coated with the liquid crystal film. The sandwich structure thus obtained is shown in Fig. 6a. This sandwich structure is now charged as illustrated in Fig. 6b. Since the liquid crystal film is not a perfect insulator, the charges migrate through the liquid crystal to selenium interface (Fig. 6c). The sandwich is now exposed as shown in Fig. 6d and positive charges on the selenium surface only remain in unexposed (dark) areas.

Under the influence of the high electrostatic fields used in the xerographic process (around 10^5 V/cm), the liquid crystal material migrates from charged to uncharged regions and may leave bare areas as illustrated in Fig. 6e.

It has been observed that the liquid crystal can be removed by the fields to the extent of leaving areas as wide as 0.3 cm completely bare. If, instead of an optical positive, a negative input is used (white image on a black background), piling up instead of excavation of the liquid crystal on the image area can be observed.

The deformation process is a function of time, and both depth and extent of

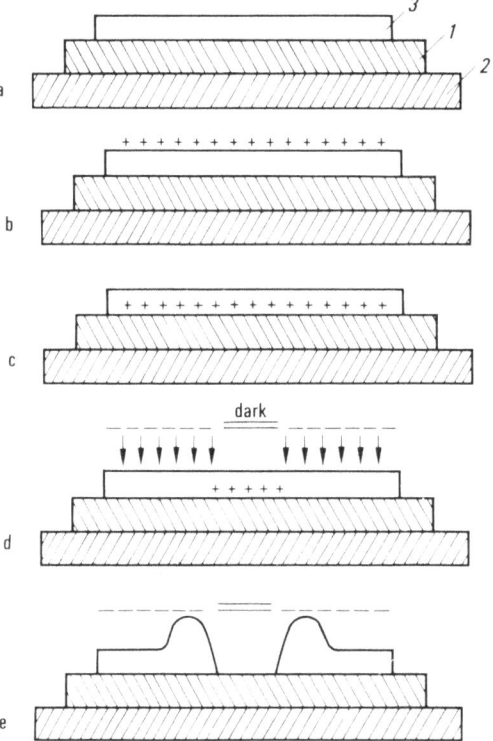

Figure 6. Electrophotographic imaging with cholesteric liquid crystals (after reference [21]).
(a) Sandwich structure used in the imaging process
(b) Sandwich structure immediately after corona charging
(c) Subsequent charge migration
(d) Exposure of target
(e) Deformation of liquid crystal film after exposure positive input
1   Selenium layer (50 μm)   3   Cholesteric liquid crystal film (1 μm – 25 μm)
2   Metal substrate

the deformation will depend on the time the liquid crystal film is subjected to the electric field. It is also a function of the viscosity of the films used; lowering the viscosity produces a more pronounced deformation.

The film thickness should be between 1 μm and 8 μm. Films thinner than 1 μm result in images mainly due to the deformation and exhibit interference colors. Films more than 8 μm thick have poor resolution, and the film surface in unexposed areas loses its uniformity and flatness. The resolution is generally a function of thickness, viscosity, brightness and contrast. Films of 3 μm thickness resolve at least 5 Lines/mm with high brightness.

2. Application of Nematic Liquid Crystals

2.1. Introduction

Conventional indicator devices are becoming increasingly less capable of meeting the more demanding requirements placed on information displays in the various branches of today's industry and business. Today the indicator device must assimilate the data rapidly and display it clearly, possibly in color. It must also consume a minimum of electric power, require only low level drive voltages and be compact. Apart from light-emitting diodes (LEDs) and plasma displays, displays with a liquid crystal layer as optically active medium meet these requirements. In thin layers nematic liquid crystals change their transmission properties for normal or polarized light when subjected to an electric field (see also first part of this book). This phenomena can be utilized for alphanumeric and analog displays, image converters and matrix-type picture screens [22, 23, 24].

Unlike conventional data display systems, a liquid crystal display radiates no light of its own, but affects incident light.

The specific advantages of liquid crystal displays are:
1. Relatively low input power per unit area. Typical values are 10 μW/cm^2 to 100 μW/cm^2 [25];
2. Viewable over a wide range of ambient lighting conditions;
3. By using suitable liquid crystal mixtures, information can be stored [26];
4. Relatively easy to design for two or more colors [27];
5. Large area display by very small volume;
6. Low power input does not require mains as an energy source;
7. Compatible with integrated circuits because of their low operating voltage;
8. Can be used in multiplex operation.

Liquid crystal displays will probably soon completely supplement existing indicator systems; further improvements can still be expected in this field [28, 29].

2.2. Liquid Crystal Display Technology

2.2.1. Basic Design

A liquid crystal display in its simplest form consists of two glass plates coated on their inner surface with an electrode layer between which a liquid crystal layer about 12 μm thick is sandwiched (Fig. 7a). A voltage applied to these electrodes produces an electric field that affects the liquid crystal. The electrode layer can be homogeneous or a pattern can be etched into it so that by driving appropriate elements various pictures can be created (Fig. 7b). To observe the electro-optic effects of the liquid crystal layer at least one of the electrode plates must be transparent. If the liquid crystal layer is operated in the transmissive mode, both electrodes must be transparent. For reflective mode operation the back electrode must be a suitable reflecting layer.

2.2.2. Fabricating the Electrode Structures

The glass plates, usually of crystal plate-glass, as substrate for the electrode structure of a liquid crystal display must meet a number of requirements: they must be

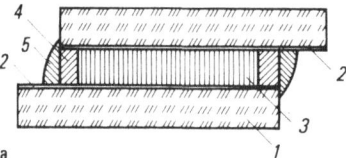

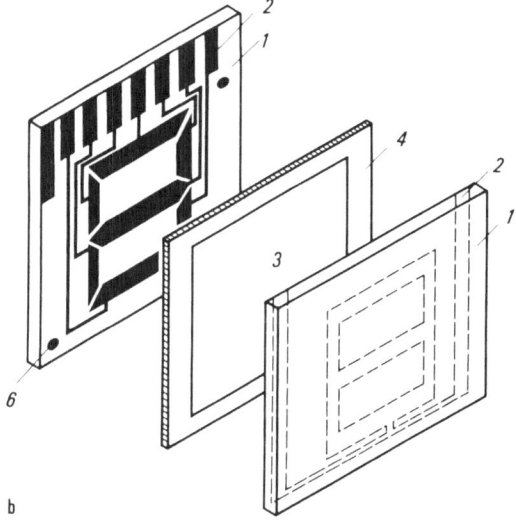

Figure 7. Liquid crystal display setup.
(a) Basic design
(b) Setup of a digital seven segment display

1	Glass plate	4	Spacing
2	Electrode layer	5	Seal
3	Liquid crystal layer	6	Filling hole

resistant to the liquid crystal material and are therefore best coated with an inert protective layer. The surface of the glass plate must be clean and completely smooth, i.e. it must be free of contamination, scratches and holes, otherwise uniform orientation of the liquid crystal layer is not possible. To achieve an even field strength and hence an uniform electro-optical effect over the entire display, the liquid crystal layer must have constant thickness. The glass substrates must therefore be very flat. For a liquid crystal layer, 12 μm thick, a tolerance of ± 3 μm is admissible for normal requirements on the display.

As transparent electrodes SnO_2 and In_2O_3 or a mixture of these two compounds are used today. The layers, 50 nm to 100 nm thick are deposited on the glass substrates either by pyrolytic, reactive coating [30] or sputtering [31, 32]. In the reactive coating process a mixture of $SnCl_4$ and water vapor in a nitrogen stream is passed over the glass heated to about 400°C. From the reaction $SnCl_4 + 2H_2O \rightarrow SnO_2 + 4HCl$ the conducting SnO_2 layer is deposited on the glass substrate. The adhesive strength and conductivity of this layer (approx. $10^2 \Omega/\square$ to $10^3 \Omega/\square$) are increased by mixing $SbCl_3$ with the $SnCl_4$. In the sputtering process the layer material in pure form as a target is vaporized by ion bombardment and deposits as a thin film on the glass substrate. This method is suitable for fabricating a layer of a single component such as SnO_2 or mixture such as SnO_2 and In_2O_3.

Vapor-deposited Al or multiple dielectric films are suitable for reflective electrically conductive layers.

A simple method of producing the electrode structure is to etch the transparent conducting layers suitably. The conducting layers on the glass substrate are first completely covered with photoresist either by spraying or by rolling on in foil form. In both cases large area layers of uniform thickness can be obtained. Exposure of the photo-resist through a mask gives the required electrode pattern on the layer. With a negative photoresist the unexposed portion is removed during the developing process. Conversely with positive photoresist the exposed portion is removed. Electrode structures with a conductor path width exceeding 50 μm, as usual for numerical displays, can be made by the screen printing process. In this case the glass substrate is partially coated with printing varnish through a mesh corresponding to the positive of the electrode structure.

After coating by the aforementioned processes the varnish layers are dried and hardened, then the free conducting layers are etched away with a suitable reagent. SnO_2-layers are dusted with Zn powder and then dipped in HCl. The reaction of HCl with Zn powder produces hydrogen in statu nascendi which reduces the SnO_2- to metallic Sn ($SnO_2 + 4H \rightarrow Sn + 2H_2O$). Sn then dissolves in HCl. In_2O_3-layers are etched with HCl, Al-layers with H_3PO_4. After etching, the varnish still on the substrate is then removed with a suitable solvent. It must be assured that the transparent glass substrate is not chemically attacked by the varnish remover.

If the electrode structure is produced by the so-called negative process, etching can be avoided. In the screen printing process the negative of the electrode structure is first deposited on the glass substrate as an easily removable layer of say MgO. After coating by one of the aforementioned methods the conductive layer is applied. The easily detachable layer is then removed, leaving the electrode structure behind on the glass substrate.

2.2.3. Production of Uniformly Aligned Liquid Crystal Layers

2.2.3.1. Surface Treatment for Liquid Crystal Alignment

Uniformly oriented, thin liquid crystal layers are required for most liquid crystal device applications, for example using dynamic scattering, deformation of aligned phases and twisted nematic layers; uniform orientation is a prerequisite for high contrast and good readability. The usual method of producing uniformly oriented thin liquid crystal layers is to generate substrate surfaces that have an orienting action on the liquid crystal molecules (see also first part of this book). Orienting surfaces can be created by various methods, including chemical treatment of the surfaces, physical adsorption, deposition of organic or inorganic thin films and surfaces with uniform microprofile. Depending on the required electrooptic effect, the homeotropic, the tilted homeotropic and inhomogeneous or homogeneous planar orientation is used.

Homeotropic orientation can be obtained by treating the substrate surfaces with acids such as chromosulphuric acid, by applying thin even monomolecular layers of lecithin [33] or dimethylpolysiloxane [34, 35], or by doping the liquid crystal with surface agents such as APAP, EBPAP or HMAB [35]. When using glass substrates, according to Uchida et al. [36], dipolar interactions between polar groups at one end of the liquid crystal molecules or surface agents and the silanol groups on the glass surface (Fig. 8) are decisive for homeotropic orientation. Also van der Waal's interactions between nonpolar alkyl-(butyl)-groups of the liquid crystal and the aligned nonpolar alkyl-groups of surface agents such as lecithin, alkoxysilane DMOAP (Fig. 9a) or dimethylpolysiloxane (Fig. 9b) give homeotropic orientation [35, 37]. Substrates coated with MgF_2, ZnS or Al_2O_3 layers also produce a homeotropic oriented liquid crystal layer [38].

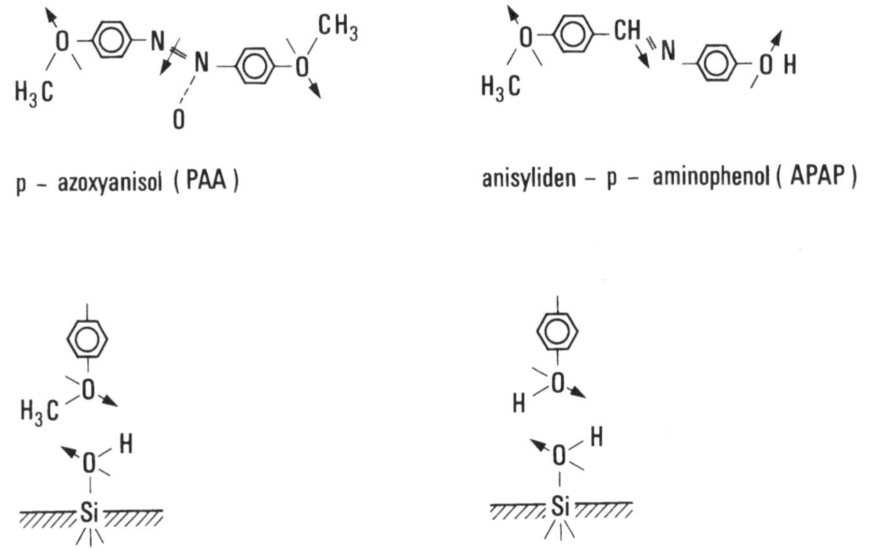

p – azoxyanisol (PAA)   anisyliden – p – aminophenol (APAP)

Figure 8. Dipolar interactions between polar groups of liquid crystals and silanol groups of glass surfaces (after reference [36]).

An inhomogeneous planar orientation is generally given on glass, SnO_2, metal or plexiglass surfaces that have not been specially treated. Kahn et al. [39] showed that a defined inhomogeneous planar orientation is given on numerous substrates when these are treated with monomolecular silane surface coupling agents where the propyl hydrocarbon chains lie randomly oriented parallel to the surface.

In many applications orientation parallel to a preferred in-plane direction (homogeneous planar orientation) is important. This can be accomplished by creating micro-grooves in the substrate (Fig. 10), for example by rubbing with a polishing cloth soaked in a suitable paste [40, 41, 42]. The quality of the homogeneous orientation depends on the topology of these microgrooves. Experiments have shown that with the nematic substances MBBA and NV a perfect homogeneous

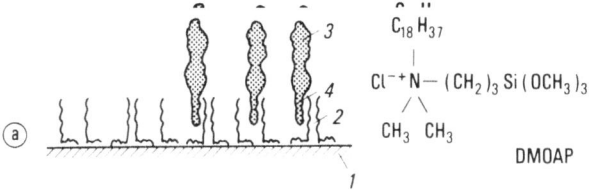

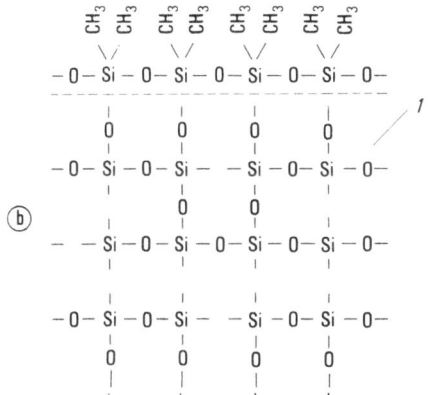

Figure 9. Homeotropic alignment by Van der Waal's interactions between nonpolar alkyl-groups of the liquid crystal molecules and the aligned nonpolar alkyl-groups of the liquid crystal molecules and the aligned nonpolar alkyl-groups of silane surface coupling agents (after reference [35]).
(a) Alignment by the alkoxysilane DMOAP
(b) Alignment by dimethylpolysiloxane $(SiO)_n \cdot (CH_3)_{2n}$

1	Glass substrate	3	Liquid crystal molecule
2	Alkyl-groups C_nH_{2n+1} of the surface coupling agent	4	Alkyl-group of the liquid crystal

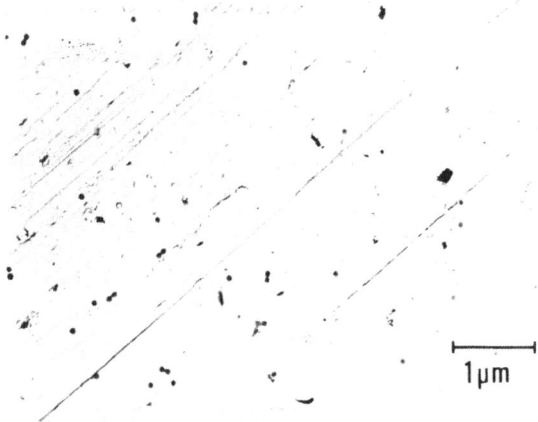

Figure 10. Electron micrograph of a platinum shadowed carbon replica of a tin oxid film rubbed with 1 μm diamond paste (after reference [41], [42]).

orientation is obtained when for a groove width of 1 μm the groove spacing is less than 10 μm and the groove depth is about 10 nm [41, 42]. Good homogeneous orientation can be obtained on substrates with electrode structures when they are first coated with an isolating layer and microgrooves are rubbed into this layer.

An elegant method of producing homogeneous alignment of nematic liquid crystals on substrates with high reproducibility is given by vapour depositing suitable materials on the substrates at a large angle to the normal. Janning [43] found that good homogeneous orientation of MBBA is supported by films of Cr, Pt, Al, Au and SiO which are less than 10 nm thick and have been evaporated onto a glass substrate using a process in which the evaporated molecules are incident at a grazing angle to the sample substrate (85° from the sample normal). Electron microscope investigations have revealed that latticed structures with a uniform direction are created by this technique [44].

2.2.3.2. Discussion of Surface Alignment

A solid surface may be characterized by a critical surface tension γ_C and a liquid by a surface tension γ_L. In general, for liquids having a surface tension γ_L, wetting of the solid occurs when

$$\gamma_L < \gamma_C.$$

In this case the liquid spreads uniformly on the solid. Conversely for

$$\gamma_L > \gamma_C$$

the liquid does not wet the surface. It exists as a sessile drop on the surface.

Creagh and Kmetz [45] have shown that substrates with γ_C less than γ_L for the liquid crystals cause homeotropic alignment of the liquid crystal molecules while $\gamma_C > \gamma_L$ results in molecular alignment parallel to the substrate. The alignment of a particular liquid crystal of known surface tension γ_C can be predicted merely by knowing the relative magnitudes of γ_C and γ_L [39]. The difference

$$\Delta\gamma = \gamma_L - \gamma_C$$

is a measure of the free energy favoring liquid crystal molecule orientation normal to a surface over orientation parallel to a surface. Creagh and Kmetz [45] observe that "when the substrate surface tension is relatively low, the intermolecular forces within the liquid crystal are stronger than the forces across the interface. The fluid does not wet the solid, and the elongated nematogenic molecules align perpendicular to the surface to maximize their intermolecular interactions. . . . When the critical surface tension of the solid is greater than that of the liquid crystal, the forces across the interface dominate. The surface energy is minimized if the fluid wets the solid and the elongated liquid crystal molecules pack flat against the substrate."

They note that adsorbed moisture from a room of normal humidity or inherent adsorbed monolayers can cause a significant reduction in γ_C. The adsorbed mono-layers may be composed of adsorbed moisture, liquid crystal molecules themselves, impurities originally found in the liquid crystals, hydrolysis products resulting from hydrolysis of the liquid crystal by surface-adsorbed moisture, and /or other reaction products resulting from interaction between liquid crystal molecules and the surface.

A typical value for γ_L of a liquid crystal is $(28.8 \pm 0.3) \cdot 10^{-3}$ N/m at 23°C. Typical values of γ_C for organic surfaces range from $6 \cdot 10^{-3}$N/m for perfluoro-lauric acid to $46 \cdot 10^{-3}$ N/m for nylon 6.6, and from $22.5 \cdot 10^{-3}$ N/m to $45 \cdot 10^{-3}$N/m for a variety of surfaces coated with organosilane coupling agents. Some additional γ_C values are $25 \cdot 10^{-3}$N/m for lecithin and $22.5 \cdot 10^{-3}$N/m for methyltrimethoxysilane. Thus although lecithin has been found to give excellent homeotropic orientation of liquid crystals, methyltrimethoxysilane might be expected to give even better homeotropic orientation because $\Delta\gamma$ would be larger for the latter surface ($6.3 \cdot 10^{-3}$N/m compared with $3.8 \cdot 10^{-3}$N/m for lecithin.)

Although dry inorganic surfaces generally have γ_C values considerably higher than those for organic surfaces, adsorbed water vapor or other contaminants reduce γ_C appreciably. For example, organic-free, sodalime glass has γ_C equal to about 75, 45, 38 and $30 \cdot 10^{-3}$N/m at relative humidities of 10^{-3}, 1, 10 and 90% respectively. Surfaces coated with organosilane coupling agents (see Section 2.2.3.1.) are exceptionally stable in the presence of moisture [46] and should therefore provide the constant values of γ_C desired for liquid crystal device structures. Chemical bonding of the organosilane molecules to the surfaces prevents adsorption of other molecules which could otherwise reduce γ_C.

In concluding surface tension, one can say that the most stable and best oriented liquid crystal cells will be those with liquid crystal-surface combinations with large $|\Delta\gamma|$. The typical γ_L and γ_C values indicate that it should usually be possible to obtain $|\Delta\gamma|$ of about $5 \cdot 10^{-3}$N/m to $15 \cdot 10^{-3}$N/m.

For parallel aligned samples with $\Delta\gamma < 0$, grooving the surface presents a good approach to obtain homogeneous nematic orientations useful in twisted nematic devices. Similarly for normal aligned samples with $\Delta\gamma > 0$, grooving the surface provides a good method of making high quality electric-field tuned birefringence devices with negative dielectric anisotropy nematic materials. It will generally be desirable to make $|\Delta\gamma|$ as large as possible in order to obtain stable liquid crystal alignment.

The following considerations were made to explain the homogeneous orientation of liquid crystal molecules: with an orientation of the molecule axes parallel to the interface, for example on glass, initially no direction is preferred. However, if the substrate surface has a wavy deformation, as caused, for example, by grooves, then the liquid crystal will be elastically deformed. The deformation energy g is given thus by [40]

$$g = 8\pi^4 \bar{K} \sin^2\theta \left(\frac{a^2}{\Lambda^4}\right) \exp(-4\pi z/\Lambda)$$

with

$\bar{K}$ mean elastic constant,

a   deformation amplitude,

Λ groove spacing,

z   coordinate normal to the surface.

The deformation energy depends on the angle θ between the director L_o of the liquid crystal and the groove direction (Fig. 11). If this angle is zero, the deformation energy disappears. However, with the preferred direction of the liquid crystals

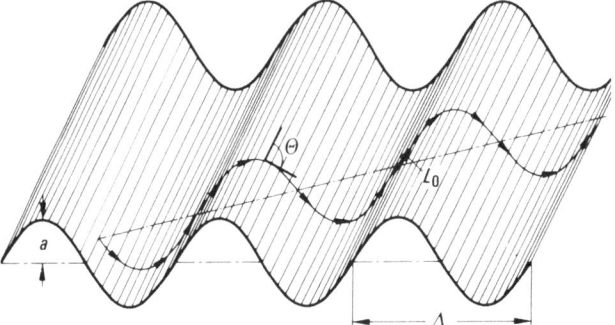

Figure 11. Sinusoidally shaped surface with a deformation amplitude α and a groove-spacing A. The direction L_0 of the liquid crystal lies in the surface as indicated by the arrows and makes an angle θ with respect to the groove direction (after reference [41], [42]).

perpendicular to the grooves, the liquid crystal deformation energy is maximum.

From these considerations it appears plausible that by rubbing the substrate, in other words producing grooves in the surface, the liquid crystal molecules can be aligned uniformly parallel to the rubbed surface and brought into a stable position. Moreover it can be concluded that only in special cases the orientation of the liquid crystal molecules normal to the grooves (maximum deformation energy of the liquid crystal) is stable [41, 42].

2.2.4. Electrode Spacing, Assembly and Filling the Display

Present conventional displays have a 10 μm to 20 μm thick liquid crystal layer sandwiched between the two electrode plates. A suitable method to measure the distance of the plates was described by Kashnow [47]. The simplest method of maintaining the spacing of the plates is to introduce foil or wires between them. Suitable foil materials are teflon or hostaphan. However, this technique is too complex for use in the production line of liquid crystal displays. Instead varnish, metal or glass layers are applied to one or both of the electrode plates along the edges by the screen printing process (Fig. 12). After hardening or stoving, the layers maintain the correct spacing between the plates. When glass layers are employed for spacing, glass balls with a higher melting point are embedded in the layers [48]. The maximum ball diameter corresponds to the thickness of the liquid crystal layer subsequently applied.

After applying the spacers, the substrates with their electrode structures are aligned and then hermetically sealed. When foil or varnish is used as spacer material, the substrates are best sealed with an adhesive such as epoxy. With metal spacers the substrates are simply soldered together. When using glass solder as spacer, the plates printed with glass solder frames are pressed together and heated.

The liquid crystal material is filled in through two holes in the electrode plates either under vacuum or by injection filling. Prior to filling it is advisable to degas the liquid crystal under vacuum and then filter it. Dopants to assure optimum

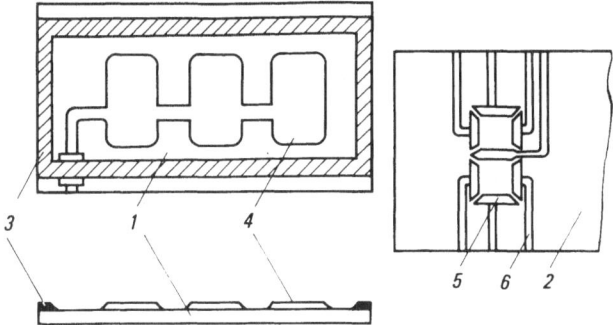

Figure 12. Setup of a reflective liquid crystal display with printed spacers.
1   Back plate   4   Printed reflective electrode
2   Face plate   5   Printed transparent electrode
3   Metallization for solder seal   6   Printed transparent conducting strip

electrical conductivity and surface agents, for example to give homeotropic orientation of the layer are mixed with the liquid crystal before filling.

2.3. Operating Modes of Liquid Crystal Displays

2.3.1. Transmissive Mode Operation

Based on experience the following rule governs the selection of the type of display if an additional light source to illuminate the display is not too complex or costly, or when maximum contrast is required, the best results are obtained with transmissive mode devices (Fig. 13a). With no additional light source available operation in the reflective mode (Fig. 13b) is indispensible.

For scattering displays operated in the transmissive mode the light source is arranged such that the observer sees a black background when the display is not activated but illuminated. The contrast with the display activated is then optimum. This configuration results in a certain depth that can be reduced by adding a light control display film from 3M[1]. The display film only allows a high percentage light transmission within a narrow viewing angle. This allows liquid crystal displays to be illuminated by diffuse light.

The display film consists of black, evenly spaced louvers a few micrometers thick implanted at a specific angle in an approx. 250 μm thick transparent film of cellulose acetate butyrate. A single layer of this display film transmits approx. 85% of the light which is incident parallel to the louver direction. For light incident from other directions the light transmission decreases sharply.

Fig. 14 shows the optimum design of a scattered light transmissive mode display with added display film. The louver angle is 40°. The light from a fluorescent lamp is collimated by two reflectors and passed in this angle parallel to the louvers

---

[1] 3M:Minnesota Mining and Manufacturing Company MBH Scotchlite-foils.

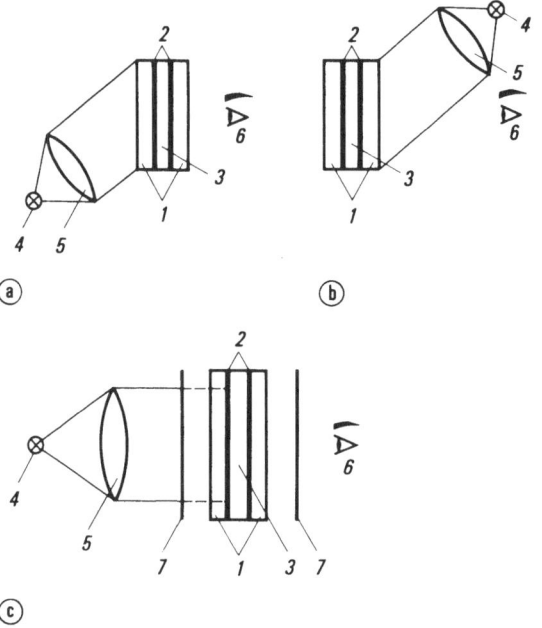

Figure 13. Illumination of liquid crystal displays.
(a) Scattering display in the transmissive mode
(b) Scattering display in the reflective mode
(c) Transmissive field effect display between crossed polarizers

1	Glass plate	4	Light source
2	Transparent electrodes	5	Lens
3	Liquid crystal		

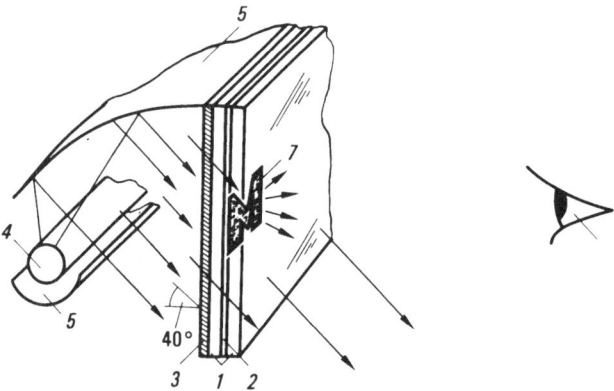

Figure 14. Scattered light transmissive mode display with light control (display) film.

1	Glass plate with electrodes	5	Reflector
2	Liquid crystal	6	Observer
3	Light control film	7	Activated liquid crystal
4	Fluorescent lamp		

through the display film. The light unaffected by the liquid crystal display is invisible to the observer.

Transmissive mode displays in which orientation effects are utilized, are located between two crossed polarizer films (Fig. 13c). The illuminating source can mostly be situated immediately behind the liquid crystal display. Colored liquid crystal displays which, for example, exploit the DAP effect (Deformation of Aligned Phases), must be illuminated by collimated light to avoid dependence of the color on the viewing angle. Moreover the front electrode must be a scattering disc. Collimated illumination is not required for color displays with homogeneously oriented liquid crystal layers, for example when the depolarization effect is used.

2.3.2. Reflective Mode Operation

Liquid crystal displays utilizing scattering effects operated in the reflective mode have a mirrored backing that throws the light scattered at the liquid crystal back through the device into the viewing field. Reflective back electrodes generally limit the readability appreciably because of unwanted reflections. Display devices of this type are therefore hardly usable in practice unless special measures are taken.

Reflective mode displays in which the back electrode is coated with a retro-reflector are free of undesired reflexes [49]. Commercially available retroreflectors in foil form exhibit the property of bending back incident light rays along the direction of incidence. They consist of a foil with embedded glass spherules about 40 μm in diameter with part of their surface mirror-coated as shown in Fig. 15a. The refractive index and diameter of the spherules are so selected that their focal point is on the surface. Even when the incident light is divergent the incident and reflected rays coincide because the ratio of light source distance to spherule diameter is mostly very large.

Fig. 15b illustrates schematically the ray path of a scatter cell backed with retroreflective foil. In the upper part of the cell the liquid crystal layer is in the scattering mode. A light ray r_1 is scattered at the liquid crystal primarily in the forward direction, reflected at the retroreflector and again scattered when passing through the cell a second time. The ray r_2 incident on a nonscattering point of the liquid crystal layer is not deflected and the incident and returning rays coincide. In principle when retroreflectors are used in liquid crystal displays, no light from non-scattering elements of the display can reach the observers eye independent of the viewing angle. Fig. 16 demonstrates the picture quality obtained in a reflective mode display, (a) backed with a retroreflector, (b) with a reflecting Al-back electrode.

Multiple dielectric layers provided with an electrode structure can also be employed as reflecting back electrodes of reflective mode displays. Dielectric layers reflect incident light like normal mirrors such as glass plates with vapor-deposited Al-film, yet they increase the contrast and hence display readability quite appreciably. They consist of a system of plane parallel, dielectric nonabsorbing layers of different refractive indices n (Fig. 17). The individual layers are approximately one light wavelength thick. When light falls on such a layer system, at each interface between adjacent layers the light is partially reflected and partially transmitted. The interference of the reflected partial waves leads, in certain spectral ranges to a

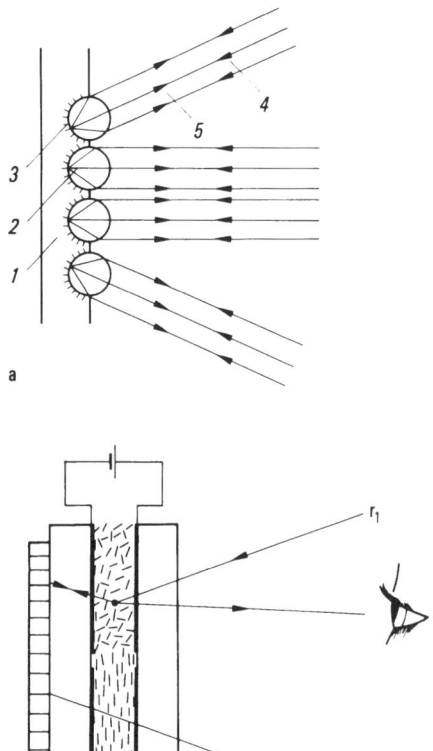

a

b

Figure 15. Reflective mode display with a retroreflector. Schematic of the ray path (after reference [49]).
(a) In the retroreflective foil
(b) In the cell

1 Supporting foil	6 Glass plates with transparent electrodes
2 Glass spherule	7 Liquid crystal
3 Mirror coated surface	8 Retroreflective foil
4 Incident light	9 Observer
5 Reflected light	

reduction in reflection, depending on the layer system configuration. With white incident light the reflected light is slightly colored. If, for example, the display is activated to dynamic scattering by applying an electric field, at the dynamic scattering points the light becomes incoherent. Hence the dynamic scattering regions of the liquid crystal appear white on an evenly colored background or in an uniformly colored surrounding. Multiple dielectric layers are, however, relatively expensive to manufacture.

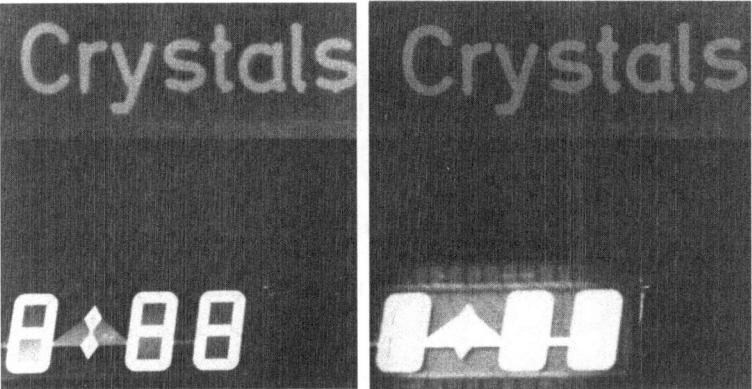

Figure 16. Comparison of reflective mode displays.
(a) With retroreflective foil
(b) With reflecting Al-backelectrode; left: Optimum illumination; right: Unfavourable illumination

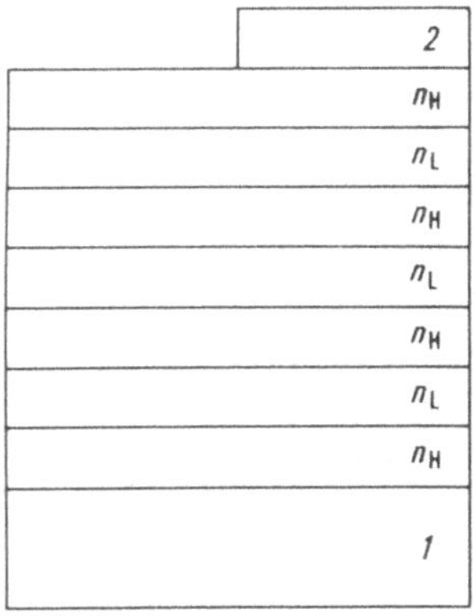

2
n_H
n_L
n_H
n_L
n_H
n_L
n_H
1

Figure 17. Schematic of a reflecting electrode consisting of a glass plate coated with multiple dielectric layers and with a conducting layer on top.
1   Glass plate
2   Conducting layer; $n_{H,L}$ high, low refractive index.

2.4. Types of Simple Liquid Crystal Displays

2.4.1. Digital or Alphanumeric Displays

The first liquid crystal displays on the market were segment devices to display numbers or letters. The most common types are 7- or 7(8) X 5-segment devices, and more seldom 14-segment displays. Fig. 18 shows the etching pattern for such displays. Depending on the type of back electrode these displays can be operated in the transmissive or reflective modes.

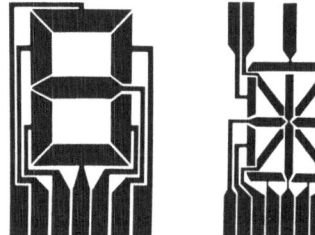 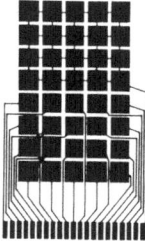

Figure 18. Etching pattern of the front electrode of 7-, 14-, 8 × 5-segment displays.

The so-called number system can only be operated in the transmissive mode. The numbers from zero to nine are arranged in sequence on a set of 11 plates. Each plate is coated on both sides with SnO_2. Apart from one plate each has a number on the back, the rest of the SnO_2 is removed. The back electrode connection of each plate is brought out separately. The front sides are connected together to form a common front electrode. This concept allows various pictures to be displayed chronologically on a relatively small surface area, making it especially attractive for advertising and direction indicator panels.

In contrast to these common liquid crystal panels, there are also panels with electrodes being arranged on one plate. The electrodes are similar to combs which penetrate each other. Spacing between the fine electrodes is about 6 μm to 25 μm. The comb-shaped electrode is said to be superior to the common one, with regard to long life operation and to be suitable for birefringence mode operation [50, 51].

Digital liquid crystal displays can be used in calculators, electronic clocks, multimeters and data output equipment [52]. With digital displays the test results are printed out, thereby eliminating reading errors. Gas discharge numeric indicator tubes require a firing voltage of about 140 V; by comparison numeric liquid crystal displays are compatible with integrated circuit voltage levels.

2.4.2. Signal Displays

There are numerous applications for liquid crystal displays in traffic and industrial control systems, particularly as switch position indicators on synoptic panels. A switch position indicator as shown in Fig. 19 functions as follows: In the normal case the four quadrants in the corners of the square display are always activated and therefore appear bright. In the event of a fault, when the ancillary voltage fails, the quadrants become transparent. To indicate whether a line is busy or free the horizontal or vertical bars appear bright. If the display contains a homogeneously or homeotropically oriented liquid crystal layer and is located between a neutral polarizer and a selective polarizer rotated by 90°, it can also display color information, for example in red. Color changeover can be achieved by adding a color filter, for example from red to green with a green filter (see also Section 2.6.4.).

2.4.3. Advertising Displays

By suitably structuring the front and back electrodes and connecting a number of

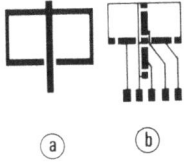

Figure 19. Switching position indicator.
(a) Back electrode
(b) Face electrode

liquid crystal displays in series, numerous different advertising displays can be created. A very simple example for using liquid crystal displays in advertising is the popular reproduction of moving phenomena. Similar to a kaleidoscope the movement is broken down into two or more phases that can be activated consecutively. Example: hopping frog, beckoning figure (Fig. 20).

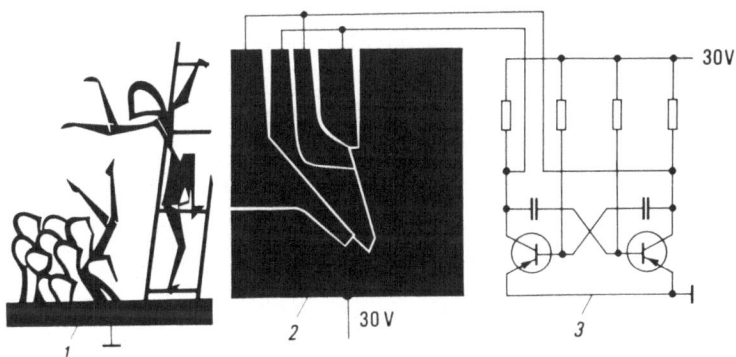

Figure 20. Example of an advertising display: Beckoning figure.
1 Face electrode
2 Back electrode
3 Electronic flipflop circuit

2.4.4. Analog Displays

By exploiting the fact that the electrical resistance of a SnO_2-layer in strip form is proportional to its length AB, liquid crystal cell displays can be used to continuously indicate measured parameters [53] (Fig. 21a, b). If the voltage U is applied to the ends of such an electrode layer, the cell appears bright where the activating threshold for the liquid crystal is exceeded. Assuming an activating threshold of 3 V and that the voltage across the cell is 6 V, a light/dark transition occurs halfway along the strip-like display. Corresponding light/dark transitions are associated with other voltages. As these transitions separate clearly defined scale ranges from each other, with the help of liquid crystals the values of physical quantities can be easily displayed by making use of the activating threshold; the light/dark transition then corresponds to a normal meter needle. Simplified versions of this display as shown in Fig. 21c, d are suitable for electrical tuning indicators. The electrode is meander- or spiral-shaped to increase the indicator resolution.

2.4.5. Image Converter

The versatility of liquid crystals as display devices is further confirmed in that an image converter can be built by simply connecting a photoconductive (CdS) layer in series with the nematic liquid crystal layer as shown in Fig. 22 [54 to 58]. The

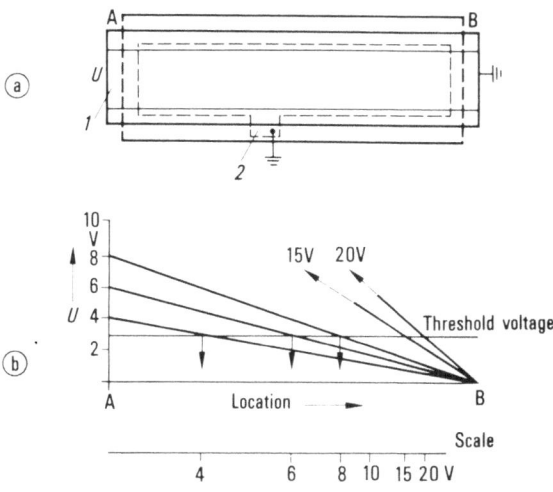

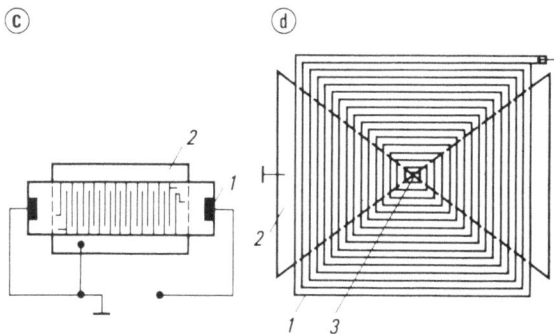

Figure 21. Analog display.
(a) Schematic of the setup
(b) Diagram for evaluating the position of the light/dark transition versus voltage
(c) Meander shaped indicator
(d) Spiral shaped indicator
1   Transparent face electrode
2   Back electrode
3   Conducting connection between face and back electrode

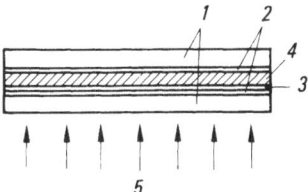

Figure 22. Schematic of an image converter with nematic liquid crystal layer.

1   Glass plate   4   Liquid crystal layer
2   Transparent electrode   5   UV-light
3   Photoconductive layer

resistance of the CdS-layer in the unilluminated case is then so high that the voltage drop across the liquid crystal layer is not sufficient to cause a visible electrooptic effect. When illuminated, the CdS-layer becomes low resistant and the liquid crystal cell is activated. Such picture screens can be used as image converters or image intensifiers especially in UV or X-ray image converters, large screen projectors and in night vision equipment.

2.4.6. Large-area Picture Screens

As liquid crystal displays do not emit light but only affect external incident light, they are characterized by very low power consumption of around 1 mW/cm^2 or less. This property makes them ideal for large-area displays. Independent of the form of electrooptic effect used — scattering, orientation, color — the liquid crystal layer in the display is generally about 10 μm thick. In the interests of a uniform effect over the entire display, this thickness must be maintained as accurately as possible. This is only assured when the substrates are extremely flat. For glass plates measuring up to 10 cm × 10 cm, the required flatness of 1 μm to 3 μm over 10 cm can be relatively easily accomplished. The requirements on larger glass plates increase proportionately. Moreover, the glass must be rigid enough not to bend. This means the glass plate thickness increases with surface area.

Large liquid crystal picture screens can be built up from rows of smaller display elements. Fig. 23a shows a large-area line of characters formed by a row of 7 × 5 matrices. Such a display is relatively expensive.

A simpler method of building up large-area liquid crystal picture screens is to replace one of the two glass plates with a foil. This eliminates the problem of substrate flatness and stability. A laboratory prototype measuring 20 cm × 20 cm has been made up by photoresisting a spacing grid on 1 mm centres with ridges 10 μm high and 100 μm wide on a glass plate coated with transparent SnO$_2$ (Fig. 23b). The recesses on this half of the display are then filled with liquid crystal. Then a hostaphan foil coated with Al for reflective mode operation or In$_2$O$_3$ for transmissive mode operation is rolled on. Finally the display is glued liquid-tight around the circumference. Experimental displays of this type have functioned well for some time, and they demonstrate the basic feasibility of a large-area liquid crystal picture screen using foils. However, the following problems became apparent:

the foil must not react with the liquid crystal. It must therefore not contain any low-molecular components such as plasticizers, lubricants or stabilizers. The foil must also be impervious, especially to liquid crystal materials and water vapor. The metallic or SnO_2 coating must be durable.

As available foils have low resistance to liquid crystal materials, attempts have been made to use glass plates or glass foil for large displays. The following technique has so far produced positive results: glass solder layers with embedded glass grains having a higher melting point are applied by the screen printing process in a dot matrix on one of the two electrode plates. The maximum diameter of the glass grains determines the thickness of the liquid crystal layer. The two glass plates are then subjected to heat treatment to space and seal them.

Unlike other types of display, liquid crystal displays can be projected like slides (Fig. 23c), with electronically controlled picture content.

a
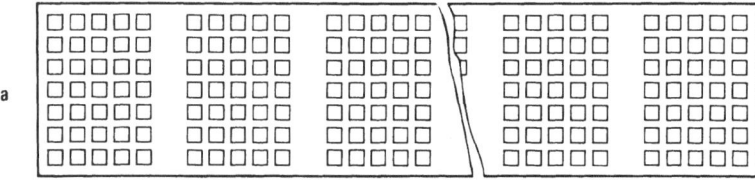

b

c
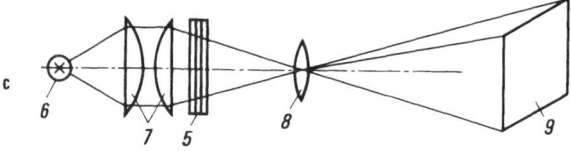

Figure 23. Large-area picture screen.
(a) Built up from smaller display elements
(b) Built up from foils
(c) Made by projection of small displays

1	Glass foil with transparent electrodes	6	Light source
2	Al foil	7	Collimator
3	Liquid crystal	8	Objective
4	Photoresist spacer	9	Screen
5	Liquid crystal display		

2.5. Electrical Characteristics of Nematic Liquid Crystal Displays

2.5.1. Characteristics of Liquid Crystal Displays

Presently available liquid crystal displays are primarily of the dynamic scattering type or of the twisted nematic type [59]. The following discusses the most important technical data of liquid crystal displays [60].

If a drive voltage is applied to a liquid crystal display and the light from the display is measured with a photocell, the intensity characteristic shown in Fig. 24 is obtained.

The time for the light intensity detected by the photocell to reach 10% of its maximum value is known as the turn-on delay. The rise time is that time required for the light intensity to increase from 10% to 90% of its maximum value. It is a function of the drive voltage and ambient temperature. It increases sharply as the drive voltage is reduced, and tends towards infinity at a certain threshold voltage.

The turn-off delay is that time required once the drive voltage is removed for

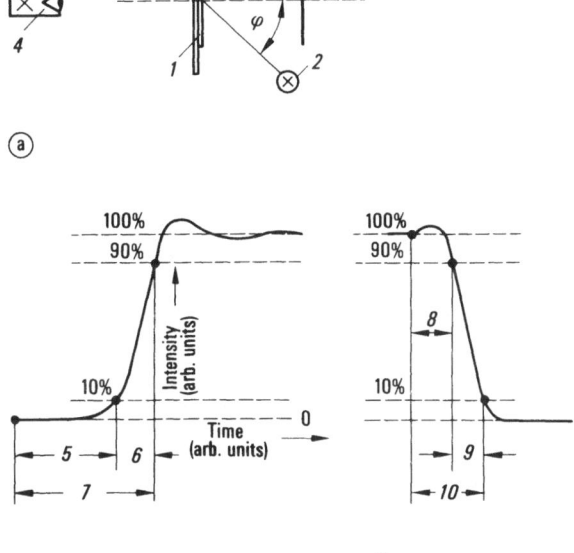

Figure 24. Schematic of the turn-on and turn-off characteristic of liquid crystal cells.
(a) Test assembly
(b) Turn-on characteristic
(c) Turn-off characteristic

1	Liquid crystal display	6 Rise time
2	Light source	7 Turn-on time
3	Black screen	8 Turn-off delay
4	Photodetector or observer	9 Fall time
5	Turn-on delay	10 Turn-off time

the light intensity to fall to 90% of its maximum value. The decay time is defined
as the time in which the light intensity falls from 90% to 10% of its maximum value.
It also depends on the ambient temperature, but unlike the rise time is hardly
dependent on the drive voltage.

In dynamic scattering displays the turn-on delay is the time required for the
hydrodynamic instability to build up. Experiments show that the hydrodynamic
instability first occurs when charge carriers have covered a distance corresponding
to the layer thickness of the liquid crystal cell. The turn-off characteristic of
dynamic scattering devices is determined by two factors:
a) turbulent flow ceases because no exciting energy is present,
b) the strongly deformed layer reorientates itself.

In field effect display cells (see also first part of this book), because there is
practically no flow of charge carriers during turn-off, only phenomenon b) occurs.
How the reorientation of the liquid crystal layer in the case of a twisted nematic
cell as a prototype of a field effect display looks like is demonstrated in Fig. 25.
Using a liquid crystal material with positive dielectric anisotropy the application of
a voltage sufficiently far above the threshold produces alignment parallel to the
electric field (Fig. 25a); if the voltage is switched off the origin spiral configuration
returns (Fig. 25b).

Characteristic data of dynamic scattering and twisted nematic displays are listed
in Table 1.

For some applications the decay times of about 100 ms are relatively long. For
a given liquid crystal material the decay time can be shortened by reducing the layer
thickness. Furthermore dynamic scattering devices with a homogeneously oriented
liquid crystal layer exhibit an appreciably shorter decay time than those with a
homeotropically oriented liquid crystal layer.

The optical quality of a liquid crystal display is best described by the contrast
that can be achieved. The contrast is defined as the ratio of light intensity with

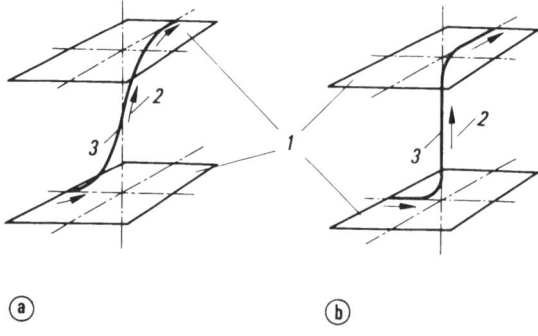

Figure 25. Orientation pattern in a twisted nematic liquid crystal layer.
(a) 90° twist, V = 0
(b) Splay and bend in a strong electric field $\vec{F}$ $(V \gg V_{th})$
1   Boundary surface
2   Director L_o of the liquid crystal
3   Orientation line of the liquid crystal

Table 1. Characteristic data of liquid crystal displays

	4-Digit 7-Segment Dynamic scattering display (Siemens)	4-1/2-Digit Twisted nematic display (Brown Bovery)
Operating voltage	25 V	3 V
Frequency	50 Hz	30 Hz to 1,5 kHz
Operating temperature	$-5°$C to $+65°$C	$-15°$C to $+60°$C
Current per segment	$2 \mu A$	$0,2 \mu A$
Capacity per segment	25 pF	60 pF
Total current	$60 \mu A$	$6 \mu A$
Total capacity	750 pF	2 nF
Rise time	25 ms	100 ms
Turn on delay	25 ms	100 ms
Fall time	80 ms	200 ms
Contrast	20:1	50:1
Storage temperature	$-20°$C to $+80°$C	

applied drive voltage to the light intensity in the quiescent state.

Also important for a display is the angle within which it can be viewed. The viewing angle of dynamic scattering devices and twisted nematic devices spreads with increasing drive voltage. It can, for example, be defined as the viewing angle within which the contrast ratio exceeds say 4:1.

2.5.2. Operating Liquid Crystal Displays by Integrated Circuits

Present liquid crystal displays require drive voltages up to 30 V depending on the type: MOS integrated circuits are therefore quite suitable to drive them. As liquid crystal displays also consume very little power, it follows that power-saving driver circuits should also be used [61, 62].

In the interest of long lifetime, AC excitation is to be preferred. Under DC excitation a form of electrolytic action occurs which causes degradation in contrast ratio during the device lifetime. So-called phase shift circuits (Fig. 26a) are ideal for AC excitation of liquid crystal displays. As Fig. 26b shows, a 50% duty cycle square wave voltage of e.g. 25 V is applied to the back electrode. The segments to be displayed are driven by antiphase voltages of the same amplitude and frequency. Unwanted segments are driven in phase with the back electrode. The difference voltage at the liquid crystal segment is then either 50 V_{pp} or 0 V. The advantage of this drive principle is that only a unipolar square wave voltage source and selector switch for only one direction of current flow are required. The phase shift is achieved with exclusive OR gates.

Low power CMOS and ion-implanted p-MOS driver circuits are presently available. Table 2 lists MOS integrated circuits specifically designed for liquid crystal displays.

If the information to be displayed is presented staggered with respect to time in BCD code, which is commonly known as multiplexing, the drive circuit shown in Fig. 27 is best used. Such a circuit contains for instance the Solitron integrated circuit type CM4117 for a $3\frac{1}{2}$-digit display. The circuit can be adapted for any

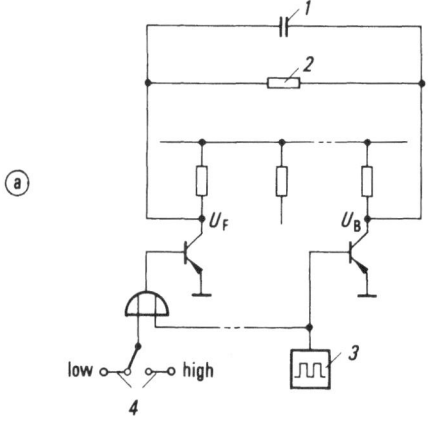

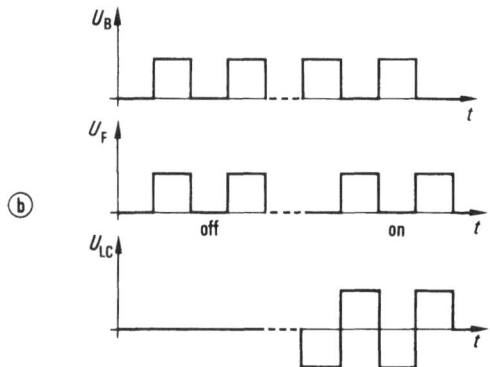

Figure 26. Operating liquid crystal display with phase shift circuit.
(a) Phase shift circuit
(b) Pulse diagram
U_F Voltage applied to face electrode
U_B Voltage applied to back electrode
U_{LC} Voltage applied to liquid crystal display
t   Time

1 Capacity of liquid crystal	3 Oscillator 50 Hz
2 **Resistance of liquid crystal**	4 Input for information

number of digits by using the Siemens integrated circuit type GDL 121 with an additional decoder and oscillator.

For data input in fixed cycle operation the AMI and Siemens package type 1907 shown in Fig. 28 has been designed. With this device and a few additional components a pulse counter, frequency meter, chronometer or digital voltmeter can be built.

Of particular interest in conjunction with liquid crystal displays are watch driver modules on a single chip that are fully electronic and require little power.

Table 2. Available Drive Packages for Liquid Crystal Displays (status Nov. 1973)

Type No.	Manufacturer	Function	Technology	Power input
1-digit 7-segment displays				
GDL101	Siemens	Counter input	p-MOS	300 mW
SFF5100	Sescosem	Counter input	p-MOS	300 mW
CD4055/56	RCA	BCD input	COSMOS	10 μW
General applications				
GDL121	Siemens	10-channel driver with memory	p-MOS	40 mW
SFF5110	Sescosem	10-channel driver with memory	p-MOS	40 mW
CD4030	RCA	4-channel driver	COSMOS	10 μW
CD4054	RCA	as CD4030 with memory	COSMOS	10 μW
Multidigit 7-segment displays				
1907	Siemens AMI	$4\frac{1}{2}$-digit counter input	p-MOS impl.	10 mW
CM4117	Solitron	$3\frac{1}{2}$-digit BCD input (multiplex)	CMOS	
Watch packages				
1998	Siemens AMI	50 Hz operation	p-MOS	400 mW
MM5316	NSC	50 Hz operation	p-MOS	400 mW
SCL5424 SCL5428 SCL5434	Solid State Scientific	32768 Hz quartz (divider SCL5423)	CMOS	10 μW
S1400	AMI	32 kHz quartz (incl. divider)	CMOS	10 μW

2.5.3. Power Consumption of Liquid Crystal Displays and MOS Driver Packages

The power consumption of a liquid crystal display device depends on the surface area of its active part. Dynamic scattering devices consume about 500 μW/cm^2. Accordingly present 4-digit displays require about 1.5 mW when all segments are activated. Field effect displays that make use of the Schadt-Helfrich effect for example require a lower operating voltage and consume less current than scattering devices. The typical power consumption of a field effect display is about 10 μW/cm^2. The ion-implanted p-MOS driver integrated circuits for a 4-digit display consume some 10 mW to 20 mW, whereas comparable CMOS packages require only about 10 μW. When extremely low power consumption is required for a liquid crystal display, as in watches, field effect displays with CMOS drivers should be preferred. For most battery-operated and mains-operated equipment, dynamic scattering devices with p-MOS drivers suffice. It is worthwhile noting that by comparison a 4-digit 7-segment display with light-emitting diodes consumes about 550 mW.

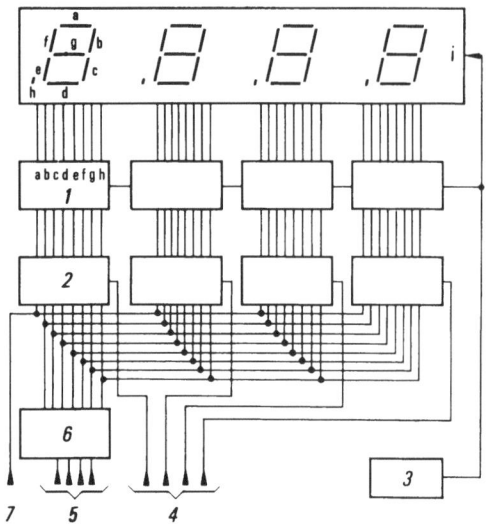

Figure 27. Addressing circuit for a 4-digit 7-segment liquid crystal display.
a to h Face electrodes
i Back electrode

1 Driver
2 8-Bit-memory
3 Oscillator 50 Hz
4 Memory address

5 Input for BCD information
6 Decoder
7 Decimal point

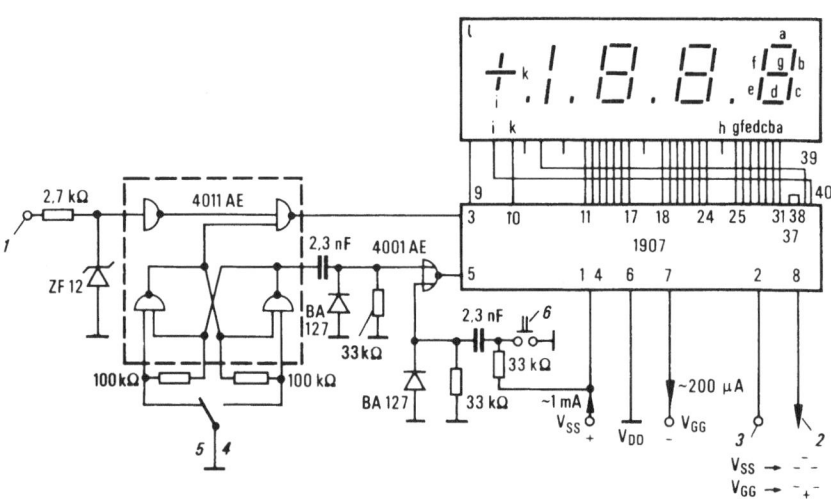

Figure 28. Digital counting circuit with drive circuit 1907 and liquid crystal display.
a to k Face electrodes
l Back electrode

1 Input
2 Over flow out
3 Polarity in

4 Stop and take over
5 Counting

2.5.4. Multiplexing Liquid Crystal Displays

With the driver circuits discussed in the preceding section a separate lead is required between each segment of the display and the corresponding output stage of the integrated circuit. A 7-segment display with more than 4 digits precludes the use of only a single integrated circuit to drive the display because various economic and technical reasons limit the number of pins on present integrated circuits to 40 or 42. The number of leads between the display and driver can be appreciably reduced by multiplex operation. With this form of drive all identical segments in each digit are connected internally to a common line; also each digit has its own back electrode. Thus for a n-digit 7-segment display, Fig. 29, only 7 + n leads are required. An 8-digit display therefore has only 15 leads compared with 57 for an unmultiplexed display.

In multiplex operation the display is driven in steps digit by digit. The associated back electrodes of each digit are driven sequentially. Together with the back electrode of each digit, the respective segments to be displayed are also driven. The circuit of such a multiplex display corresponds to a liquid crystal matrix with 7 row and n column lines. The display segments represent the matrix intersection points.

The problems associated with multiplexing are cross talk between the driven and undriven segments and reduced contrast ratio. The latter results from the fact that the scattering segments are no longer driven statically but only have voltage applied to them during a certain time interval determined by the multiplexing ratio.

No more than 4 digits can be multiplexed over a wide temperature range with present liquid crystals.

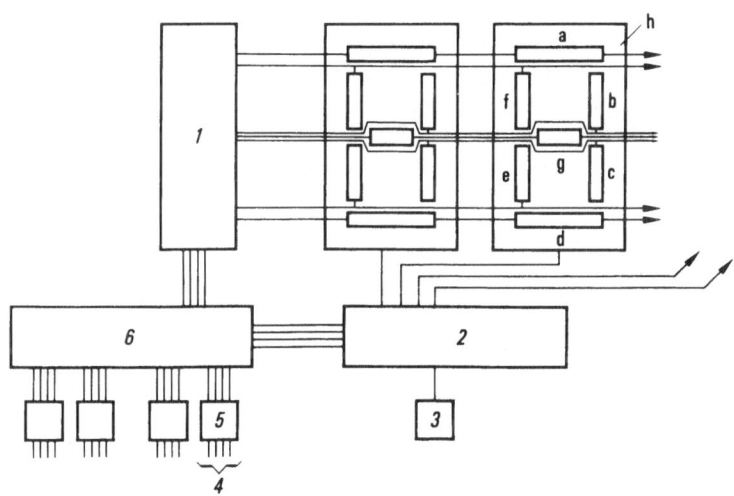

Figure 29. Diagram of a multiplex addressing for a 7-segment display.
a to g Face electrodes
h Back electrode

1 Decoder driver	4 Input for information
2 Counter driver	5 Memory
3 Oscillator	6 Multiplexer

Liquid crystals, especially those with storage properties, are now being developed which are expected to permit multiplexing of more digits. With such materials available, multidigit 7-segment, 5 × 7 dot and matrix displays would be economically justifiable.

2.5.5. Liquid Crystal Materials for Displays

2.5.5.1. Molecular Structure of Nematic Compounds

More recently especially Brown and Shaw [63], Dave and Patel [64], Gray [65, 66], de Jeu et al. [67], Kast [68] and Wiegand [69] have done much work in the field of liquid crystalline phases and have investigated the relationship between the molecular structure of organic compounds and the occurrence of anisotropic mesophases. Survey reports concerning this theme were given especially by L. Creagh [70] and R. Steinsträsser and Pohl [71]. Kast [72] listed all compounds known up to 1959 with liquid crystalline properties.

The mesomorphic-isotropic transition temperature is a guide for characterizing the thermodynamic stability of liquid crystalline phases. This is the temperature at which the anisotropic phase and the isotropic liquid phase are in equilibrium. The greater the anisotropy of the molecule polarizability, the higher the mesomorphic–isotropic transition temperature and more stable mesophases can then be expected. Accordingly, liquid crystals are generally compounds with an elongated, planar center part having a rigid molecular axis. The typical structure of liquid crystals is given by the following formula:

Concerning their applicability, compounds of this type were investigated intensively [73, 74, 75]. The center group X of the molecule induces anisotropic melt characteristics. The type and chain length of the end groups E and E′ determine which mesophase the compound exhibits. Possibilities for the center group X and the end groups E and E′ are listed in Table 3.

If the end groups of ether or ester are linked by oxygen to the center part of the molecule, the free electron pairs make an additional contribution to the π-electron system, which shows itself in thermally more stable mesophases.

Table 3. Typical Center Groups X and End Groups E and E′ (E ≠ E′) of Nematic Liquid Crystals

Center group X	Compound with center group X	End group E and E′	
$-CH=CH-$	Trans-Stilbene	$C_nH_{2n+1}-$	Alkyl
$-C=C-$	Tolane	$C_nH_{2n+1}O-$	Alkoxy
$-N=N-$	Azobenzene	$C_nH_{2n+1}COO-$	Carboxyl
$-N=N-$	Azoxybenzene	$C_nH_{2n+1}OCOO-$	Carbonic
O			Acid
			Ester group
$-CH-N(O)-$	Nitrons		
$-CH=N-$	Schiff bases		
$-O-CO-$	Phenylbenzoate		

2.5.5.2. Nematic Liquid Crystals with Special Properties

For any of the known electrooptic effects to be truly useful, the operating temperature range of the display must be sufficiently wide. Table 4 lists the requirements for various applications.

Table 4. Temperature Requirements for Liquid Crystal Devices; after L. Creagh [70]

Application	Acceptable temperature range °C	
	Storage	Operating
Commercial calculators, clocks, instrument panels, multimeters, industrial controls	−45 to +70	0 to −45
Watches	0 to +55	8 to +48 wrist temp. +28
Government	−55 to +90	−55 to +70
Automotive	−55 to +100	−55 to +70

The temperature ranges over which display devices will function are limited by the mesomorphic range of the liquid crystal materials. Although external heating may be possible, the obviously better materials for displays are those that are mesomorphic under ambient conditions.

The first compound exhibiting a nematic mesophase at room temperature, the Schiff base p-n-Butyl-N-(p-methoxybenzylidene) aniline

(MBBA)

crystalline-nematic temperature 20°C, nematic-isotropic temperature 47°C, was first described in 1969 by Kelker and Scheuerle [76].

In the meantime, binary eutectic mixtures of Schiff bases have been produced, some with melting points well below 0°C [76 to 78]. Fig. 30 shows the binary phase diagram for the components
p-n-Butyl-N-(p-methoxybenzylidene) aniline: (MBBA) and
p-n-Butyl-N-(p-ethoxybenzylidene) aniline: (EBBA)
This diagram is typical for nematic liquid crystals. However, as the tendency to hydrolysis of the Schiff bases (Section 2.5.6.1.) seriously limits their use technically, an intramolecular hydrogen bridge is introduced in an attempt to stabilize them. The stabilization improves distinctly but is still inadequate.

Investigations on stable structure types have shown that the substituent combinations p-n-Butyl/p-methoxy lead also with the Azobenzenes resp. Azoxybenzenes to compounds (a) resp. (b)

(a)

(b)

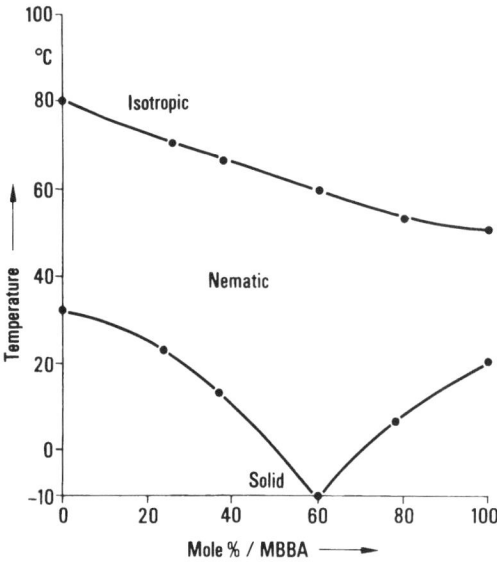

Figure 30. Binary phase diagram for MBBA and EBBA (after D. L. Fishel et al.: J. Chem. Soc. 23, 1557 (1971)).

with the lowest melting points of this class of material:
(a) crystalline-nematic temperature 32°C, nematic-isotropic temperature 47°C; resp.
(b) crystalline-nematic temperature 16°C, nematic-isotropic temperature 76°C [79, 80].

The Azoxy-product (b) is a mixture of the two N−O-isomers. By mixing with 35% of the isomeric pair p-Ethyl-p-methoxyazoxybenzene an eutectic melting at −5°C (nematic-isotropic temperature 75°C) could be obtained.

The nematic Azomethines and Azoxybenzene derivatives exhibit a negative dielectric anisotropy. If there is no permanent electric dipole perpendicular to the long axis of the molecule ($\epsilon_\perp = 0$), or one of the p-positions carries a strong polar group, the dielectric constant $\epsilon_\parallel$ in the direction of the long axis dominates and the dielectric anisotropy ($\Delta\epsilon = \epsilon_\parallel - \epsilon_\perp$) becomes positive. The dielectric anisotropy of the Azoxybenzene derivative (b) is at −0.2, whereas that of the Azobenzene analogon (a) is at +0.2.

Nitrile groups in one of the p-positions increase the dielectric anisotropy. Recently a dielectric of +14 was described for a binary eutectic system with nematic meso-phase at room temperature consisting of 2/3 (c) and 1/3 (d) with a crystalline-nematic temp. −30°C, nematic-isotropic temperature 62°C [81] :

$$n-C_6H_{13}-\bigcirc-CH=N-\bigcirc-CN \qquad (c)$$

$$n-C_3H_7-\bigcirc-CH=N-\bigcirc-CN \qquad (d)$$

What is the probability that nematic liquid crystals will become available which meet the severest requirements of any user [82] ? The lower temperature limit is more difficult to meet than the upper temperature limit. Today liquid crystals are available with a lower temperature between $-20°C$ and $0°C$ and an upper temperature between $60°C$ and $100°C$.

As far as storage at the upper temperature limit is concerned, it appears that exceeding the nematic-isotropic transition temperature is a packaging problem rather than a liquid crystal problem. Pure MBBA and pure EBBA can be contained in glass under vacuum with no change in chemical characteristics after 13,000 hours at $85°C$. Other types of liquid crystals, being more stable than Schiff bases, should behave at least as well. If package integrity is assumed, cycling the liquid crystal device through the nematic-isotropic point generally does not destroy aligning forces so that the original alignment is preserved on cooling.

The use of bulk surfactants complicates the chemical picture for storage at high temperatures. Preferential adsorption of the surfactant to the substrate may be changed by increasing temperature. Chemical reactions may occur between the liquid crystal and the additive, or the additive itself may be thermally unstable. The course of these complex chemical reactions can be followed by lowering the nematic-isotropic transition temperature. Experiments by L. Creagh [70] have shown that while MBBA alone is thermally stable, MBBA saturated with hexadecyltrimethylammonium bromide gradually degrades. In general, this upper transition temperature is quite sensitive to impurities.

2.5.5.3. Future Developments

The prospects for devices utilizing liquid crystal displays ultimately depend on customer acceptance and there is still much work to be done on investigating new liquid crystal materials to obtain wider operating temperature ranges, better electrooptic performance and higher stability. While the electrooptic principles involved in these liquid crystal displays are now reasonably well understood, controlling the electrical properties of liquid crystal materials is still largely empirical. Detailed investigations on the interaction of liquid crystals with surfaces, material purification and electrochemical behavior will assist the development of improved devices. In this respect, liquid crystals are presently in a position similar to that of the transistor in 1952, and we can look forward to many improvements and new applications in the next years [82].

To summarize, liquid crystal displays meet the requirements of some users today with regard to response times, temperature range and reliability. Much remains to be done to establish reliability and performance under the severe environments of military and automative applications. Major improvements in packaging, drive circuitry, appearance and reduced cost are necessary before liquid crystal displays become widely used.

2.5.6. Lifetime of Liquid Crystal Displays

The lifetime of presently available liquid crystal display devices significantly limits their usability. Only a.c. drive assures adequate display lifetimes of more than 10,000 hours whereas on d.c. excitation useful lifetimes have rarely exceeded 3000 hours. In the latter mode irreversible electrochemical reactions are responsible for the drastically reduced lifetime.

A Sussmann found that the useful lifetime of a d.c. driven cell is inversely proportional to the current density in the cell [83]. On the other hand the electrical conductivity of the liquid crystal determines the contrast of a display based on an electrohydrodynamic effect such as dynamic scattering. The conductivity can therefore not be decreased at will in favor of longer display lifetime. For optimum contrast the conductivity is about $10^{-8} (\Omega\, cm)^{-1}$ to $10^{-9} (\Omega\, cm)^{-1}$.

The nature of the electrochemical reactions in a cell, especially with d.c. drive, depends amongst other factors on the molecular constitution of the nematic substances, on impurities in the liquid crystal and on the conductivity and orientation additives [84]. Even the electrodes and substrates, their contamination and the drive voltage and its duty cycle can also affect these reactions. The primary failure mechanisms are breakdown of the surface alignment, formation of bubbles, discoloration of the liquid crystal layer or a narrowing of the liquid crystalline temperature range.

Field effect devices have a much lower conductivity of less than $10^{-10} (\Omega\, cm)^{-1}$. Also the operating voltages are much lower. Both factors contribute substantially to increased lifetime. L. Creagh [70] observed a direct correlation between the lifetime with d.c. drive and the lifetime with a.c. drive (proportionality); the proportionality constant has not yet been accurately determined.

2.5.6.1. Life Tests with a.c. Excitation

Displays with liquid crystal mixtures such as Azoxy-compounds or nematic ester of Schiff bases can be driven by a.c. voltages more than 5000 hours without significant changes in their electrooptic properties. The turn-on and turn-off times hardly change with lifetime. Whereas with Schiff bases the cell conductivity increases with lifetime, with Azoxy- compounds and nematic ester it remains practically constant. The conductivity of the liquid crystal consists of the inherent conductivity of the layer (bulk conductivity) and the conductivity resulting from the charge carriers injected at the electrodes. At higher field strengths the latter makes the major contribution to the total conductivity.

The injection of charge carriers at the electrode is caused by electrochemical oxidation or reduction of the liquid crystal and the impurities or dopants in the liquid crystal.

The electrochemical behavior of the presently most important liquid crystals — Schiff bases — has been studied by various authors [85 to 87]. It has been observed that decomposition results from traces of acid coincidentally present or from contamination of the cell walls. In the presence of H_3^+O ions, traces of water can hydrolyze Schiff bases:

The H_3^+N—⟨O⟩—E' ion forming during decomposition is electrochemically very active. Similar investigations on Azoxy-compounds and nematic esters have still to be conducted.

2.5.6.2. Life Tests with d.c. Excitation

As mentioned previously, Schiff bases are chemically relatively unstable. This is especially true of d.c. driven cells incorporating components that favor liquid crystal decomposition. Using the glass solder technique in cell manufacture substantially improves the stability of Schiff bases. The stability of Azoxy-compounds and nematic ester, of which the latter has only been available more recently, is hardly affected by the cell components.

The specific electrical conductivity of pure Azoxy-compounds lies between $10^{-10}(\Omega\,cm)^{-1}$ and $10^{-11}(\Omega\,cm)^{-1}$. Various dopants are added to the Azoxy-compound to influence the conductivity and orientation. The effect of these additives on the variation with time of the electrical conductivity, of the switching times, and of contrast can be measured. It has been found that after roughly 500 hours of operation for instance the cell conductance when measuring the d.c. current flowing through the cell decreases by one to two powers of ten. As subsequent a.c. measurements revealed, this phenomenon results from the formation of double layers with lower electrical conductivity near the electrodes. These double layers disturb the ion injection from the electrode. As a result of low current densities the dynamic scattering disappears almost entirely.

2.5.6.3. Photochemical Reactions in Liquid Crystals

Photochemical reactions are observed particularly in Azoxy-compounds such as NV and NVI from Merck. Displays originally pale yellow appear bright yellow after a few hundred hours of operation in sunlight. These chemical reactions, the mechanism of which is not yet fully understood, reduce the negative dielectric anisotropy of Azoxy-compounds and impair operating characteristics like the dielectric cutoff frequency. Schiff bases and especially nematic ester can withstand light irradiation better than Azoxy-compounds particularly in the blue and ultra-violet regions because of their lower absorption at these wavelengths.

2.6. Color Displays

2.6.1. Electrically Controlled Birefringence

In thin nematic films with a negative dielectric anisotropy ($\epsilon_\parallel - \epsilon_\perp < 0$) electrically controlled birefringence occurs when the two conditions are met that the nematic film is homeotropic and that the electric field $\vec{E}$ is applied parallel to the optical axis of the film [88 to 91]. When a cell fulfilling these requirements is placed between crossed polarizers, light transmission is not possible when the applied voltage is below the threshold and the incident light is perpendicular to the cell. Above threshold the homeotropic structure becomes deformed, due to the strong boundary forces and the elastic behavior of the liquid crystal inside the film more than on the substrate surface (Fig. 31). The reordered nematic no longer appears isotropic to normally incident light. Its optical indicatrix has been distorted and the film has become birefringent along $\vec{E}$. The birefringence and consequently the light transmission of the device increases with voltage. Taking into account the absorption and reflection losses of the polarizers, the optical intensity I emerging from the

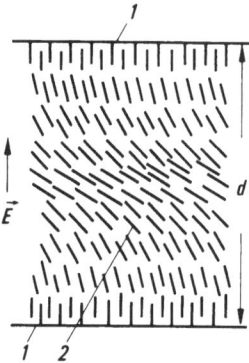

Figure 31. Deformation of a homeotropic-aligned liquid crystal film due to applied voltage.
1   Substrate surface
2   Deformed liquid crystal layer
d   distance of electrodes

aligned nematic between crossed linear polarizers – assuming normal incidence – is obtained from the formula:

$$I(\lambda) = I_p \sin^2 2\phi(U) \sin^2 (\delta/2)$$

where the relative phase retardation is $\delta = 2\pi d\Delta n/\lambda$; I_p is the transmission intensity of two parallel polarizers, ϕ is the angle between the input light polarization and the nematic optical axis (voltage-dependent), Δn is the birefringence of the film, d the film thickness and λ the optical wavelength.

The electrooptic transfer characteristic in the transmissive mode for normal incident monochromatic light and with a.c. excitation is shown in Fig. 32 for a 20 μm sample [89]. The optical transmission is expressed as a percentage of I_p.

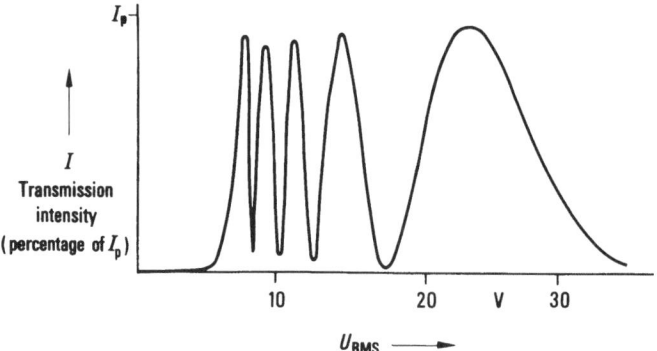

Figure 32. Electrooptic transmission characteristic of a homeotropic liquid crystal cell between crossed polarizers versus applied voltage for incident monochromatic light.
MBBA; t = 25°C; f = 1 kHz; λ = 632,8 nm; d = 20 μm (after reference [89]).

Compared to the 20 μm sample, devices with a thinner film exhibit a broader turn-on threshold.

Above threshold the light valve has a remarkable voltage-dependent spectral passband; it is in fact a tunable optical filter. According to the equation for the optical phase difference with normal incidence $\phi = 2\pi d/\lambda \Delta n$ (U), with Δn as a function of the applied voltage, each value of phase difference between zero and a certain maximum value can be set up. Extinction is obtained for a given wavelength λ for $\phi = 0, 2\pi, 4\pi, \ldots$. When using white light, spectral light attenuation is obtained which leads to the well-known Newton colors. Fig. 33 illustrates the relationship of the phase difference and the resulting color to the normalized voltage U/U_o, where U_o is the threshold voltage.

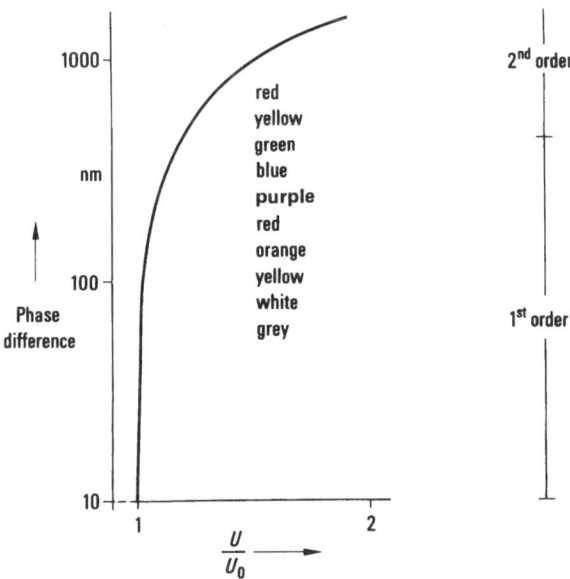

Figure 33. Relationship of the phase difference and the resulting color to the normalized voltage U/U_o, d = 20 μm; f = 5 kHz, t = 25° C (after reference [88]).

2.6.2. Electrically Controllable Domains

In sandwich cells with thin layers (thickness $\leqslant 6 \mu$m) of highly purified nematic with a resistivity of about 5×10^{10} Ω cm it has been observed [92] under the polarizing microscope that at 6 V to 10 V domain formation occurs and the domain width steadily decreases with increasing field strength (Fig. 34). The domain pattern is completely stationary at fixed voltage and is reproducible with voltage. As the domain width decreases, the total number of domains will increase. The new domains are generated between those already existing; the domains are best visible when the incident light is polarized perpendicular to the long axis of the domains and vanish with parallel polarized light.

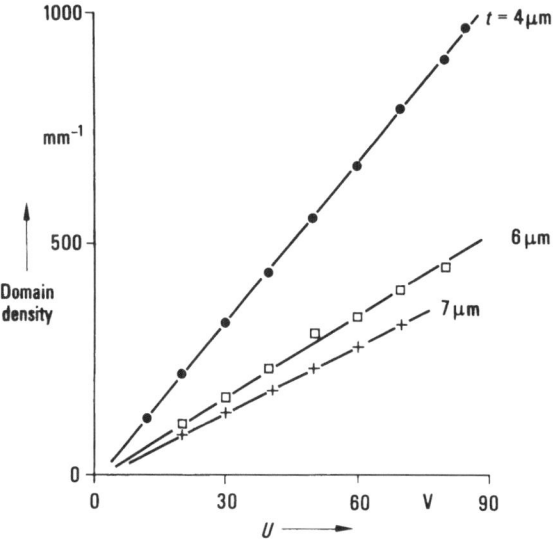

Figure 34. Domain density in a liquid crystal layer versus voltage U; t layer thickness (after reference [92]).

In white light the cell exhibiting stable domains and representing an optical grating has intense brilliant diffraction colors that can be controlled by the electric field.

Fig. 35 shows a suitable setup with a liquid crystal display for producing electrically controllable colors: With a transmissive mode display, inclined incident collimated white light is diffracted into the spectral colors when a display element is excited. A filter in front of the display only permits transmission of perpendicular incident light.

Varying the spacing of the domains of the grating structure in the activated element by changing the applied voltage allows only the perpendicular incident color of the spectrum to pass through the filter. As red light is diffracted more than

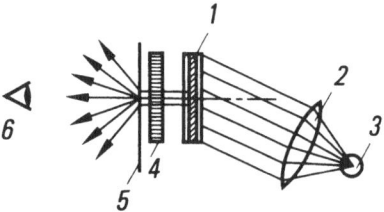

Figure 35. Setup of a liquid crystal display for producing electrically controllable colors using electrically controllable domains (after reference [92]).

1	Liquid crystal display	4	Light control film
2	Collimator	5	Diffusing screen
3	Light source	6	Observer

blue light, by applying low voltages the display appears red (large domain spacing), and by applying higher voltages the display appears blue (smaller domain spacing). An additional thin diffusing screen is used for wide angle viewing. As filter a 3M light control display film (Section 2.3.1.) has been used successfully.

2.6.3. Displays Using Guest-host Interactions

The ability of a liquid crystal to align cooperatively in an electric field can be used to orient "guest" pleochroic dye molecules. As the absorption spectrum of these molecules is a function of the molecular orientation with respect to the polarization of the incident light, one can electrically switch the color of transmitted light. Using a nematic liquid crystal with positive dielectric anisotropy and for instance methyl-red as the guest in a solution containing 0.1% to 1% by weight, a voltage applied to a thin layer changes the color of the solution from reddish-orange to yellow when viewed in transmission using polarized white light [93]. With no field the liquid crystal and pleochroic dye molecules align randomly; the dye absorbs incident light and colors the cell. An applied voltage lines up the liquid crystal and dye; light is not absorbed and the cell assumes the color of the light (Fig. 36). When the field is removed, the sample reverts back to its initial condition.

The degree of color switching is a function of the field strength. High color contrast is obtained when pleochroic dyes are dissolved in uniformly oriented liquid crystal layers. Guest-host interaction is also possible in liquid crystal layers with negative dielectric anisotropy and originally homeotropic orientation; it can also be applied to cholesteric focalconic layers using cholesteric-nematic phase transition.

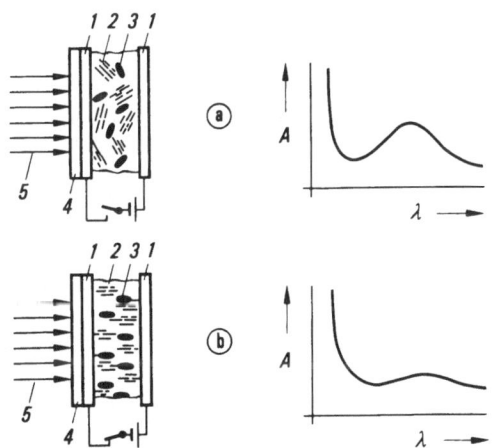

Figure 36. Electrical color switching in liquid crystal cells with pleochroic dye molecules embedded in the liquid crystal layer (after reference [93]).
(a) Absorption by random alignment of liquid crystal and dye
(b) Absorption by aligned liquid crystal and dye

1	Transparent electrode	5 Incident light
2	Nematic domains	A Absorption
3	Pleochroic dye molecule	λ Wavelength of incident light
4	Polarizer	

Measurements of optical density, or relative absorbance, as a function of applied field strength and dye concentrations taken on cells prepared with a wide variety of mixtures of dyes and nematic liquids show that contrast can be optimized by keeping the dye concentration at or below 1%. This level appears to yield the greatest absorption change. The existence of a level indicates that nematic compounds can produce alignment of only a limited number of dye molecules. As the number of dissolved dye molecules increases, non-oriented molecules that contribute to absorption of the medium under an electric field reduce contrast.

The ability to "tune" the transmission characteristics of a medium electrically using low voltages may find application as a light valve.

2.6.4. Color Displays with a Selective Polarizer

Polarization films are fabricated by embedding dichroic dyes in plastic. A special process is employed to obtain uniform alignment of the dye molecules in the substrate material. The film then only absorbs light with a defined direction of oscillation. If the dye is black the polarizer is "neutral", with any other dye the polarizer is "selective". If for instance a red dye is embedded in the film, the filter is transparent to red light of both directions of oscillation; the complementary spectral range is polarized.

When white light passes through a neutral polarizer, a homeotropic liquid crystal layer and a selective polarizer rotated at 90° to the neutral polarizer, it emerges colored, in this case red because of the transparency of the selective polarizer to red light of both directions of oscillation. If the liquid crystal cell is switched to the light scattering depolarizing mode, the arrangement transmits white light [94].

If an additional color filter is used, for example green, in the unactivated state a red color is obtained, resulting from the red component of the light transmitted through the green filter. With the liquid crystal operating in the scattering mode the color is determined by the color filter — green —. Fig. 37 shows the display construction with the liquid crystal oriented perpendicular to the layer plane when no field is applied.

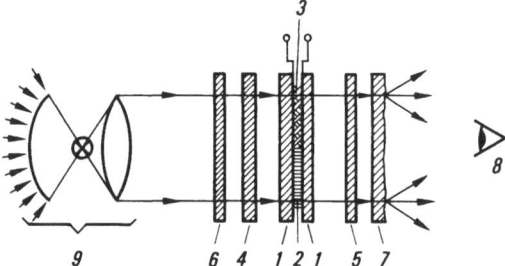

Figure 37. Setup of a color display with a selective polarizer and a color filter (after reference [94]).

1 Transparent electrodes	5 Selective polarizer
2 Homeotropic aligned liquid crystal with no field	6 Color filter
3 Light scattering depolarizing activated liquid crystal	7 Scattering disc
	8 Observer
4 Neutral polarizer	9 Light source

If it is desired to view a color display having a selective polarizer in a wide viewing range and without a scattering disc, homogeneously oriented liquid crystal layers are used. Birefringence of the liquid crystal layer is avoided when the preferred direction of the liquid crystal layer coincides with the blocking direction of the polarizer.

2.6.5. Color Displays with Retarding Layers

An optical retarding layer has different refractive indices for various directions of polarization of the light. When placed between crossed polarizers it causes color brightening. These interference colors occurring with birefringence are determined by the position of the main axes of the layer relative to the direction of polarization of the light and the retardation Δnd, where Δn is the difference in main refractive indices and d is the thickness of the layer. The brightest colors are obtained for

$$300 \text{ nm} \leqslant \Delta\text{nd} \leqslant 1500 \text{ nm}.$$

If in addition to a birefringent retarding layer a homeotropically oriented liquid crystal cell is placed between the polarizers, initially the liquid crystal layer is neutral and will not affect the light color. Applying an electric field activates the liquid crystal layer to domain formation or to dynamic scattering; it then depolarizes incident light. As retarding layers are ineffective in depolarized light, the color vanishes and the display appears white. Thus it is possible to switch from a color to white with an electric field [94].

A dual color display is obtained simply by introducing an additional color filter as shown in Fig. 38. This filter determines the display color with the liquid crystal acting as a depolarizer. If the liquid crystal layer is homeotropically oriented this color is mixed with the interference color produced by the retarding film. Any desired color combination can thus be obtained by selecting suitable color filters and retarding films. As the interference color of a birefringent layer depends on the direction of incidence of the light, color displays with retarding layers must be

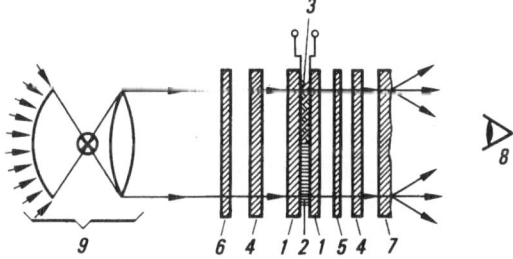

Figure 38. Setup of a color display with retarding foil (after reference [94]).

1	Transparent electrode	5	Retarding layer
2	Homeotropic aligned unactivated liquid crystal	6	Color filter
		7	Scattering disc
3	Light scattering depolarizing activated liquid crystal	8	Observer
		9	Collimated light source
4	Polarizer		

illuminated with collimated light. To increase the viewing angle the liquid crystal display must be combined with a scattering disc.

It should be mentioned that selective polarizers or retarding layers can also be applied to twisted nematic devices. Kabayashi [95] has shown that with twisted nematic cells in combination with different additional selective polarizers multicolor field effect devices with over 20 different colors are available.

2.7. Comparison of the Important Electrooptical Effects Used in Liquid Crystal Displays

At the present time the predominant electrooptic effects utilized in nematic liquid crystal displays is dynamic scattering [96] and twisted nematic effects. All the display types described above can be operated both as dynamic scattering and field effect devices. Field effect devices have been termed the second generation of liquid crystal displays, implying that they will afford a dramatic improvement over dynamic scattering devices. Of the field effects presently known, three are of major interest for use in display devices: the twisted nematic effect, the field-induced birefringence and the field-induced cholesteric-nematic phase transition. Construction of devices utilizing these effects resembles that of dynamic scattering panels. Because the nematic materials generally originate from the same chemical family, temperature ranges and packaging requirements are also similar. Although electric current flow is not necessary for the operation of field effect devices, it is difficult to eliminate in practice; d.c. operation will therefore cause electrochemical degradation as in the dynamic scattering case, although presumably at a rate reduced proportionally to the current. Establishing and retaining uniform nematic alignment over the entire device area is a more critical requirement for field effect displays than with dynamic scattering devices.

Twisted nematic devices offer evolutionary improvements over dynamic scattering displays in the two critical areas of display appearance and multiplex capability. In addition, reductions in threshold voltage and current drain may be significant for battery requirements and display lifetime.

Twisted nematic displays are simpler to illuminate than dynamic scattering devices. Backlighting with high contrast is accomplished directly through a diffusing screen. Likewise the mirror in the reflective mode scattering display is supplanted by a diffuse reflector in a twisted nematic reflecting display, thereby eliminating glare and distracting reflections. Because liquid crystal materials are available with quite large positive anisotropy, mixtures can be made with threshold voltages below 1 V, although the range 2 V to 4 V is more common. A disadvantage of the twisted nematic display is that the viewing field is rather narrow in either mode of operation.

Like a twisted nematic device, an induced birefringence device is also viewed between crossed polarizers. The color observed at a given voltage varies with the liquid crystal layer thickness, viewing angle, direction of molecular alignment and temperature. Color uniformity and repeatability therefore demand very close mechanical tolerances on the electrode plate spacing and tight control of the ambient temperature. The field of view is so narrow that this display type is only

suitable primarily for projection displays [97]. Immunity to washout in bright
ambient conditions is foregone by this approach. The main advantage of the
induced birefringent display is its relatively large multiplex capability.

The field-induced cholesteric-nematic phase transition has a uniquely fast
transient response [98], but the accompanying high voltage requirement is not
compatible with integrated circuit driver voltage levels [99]. Therefore until now
this effect was of little interest for device application. Some interest in this effect
resulted from its storage capability. As will be seen later, ideas exist to apply this
effect in future research to advantage for large storage picture screens.

2.8. Matrix-type Liquid Crystal Displays

2.8.1. General Problems

It is the objective of numerous development programs to construct a matrix-type
picture screen that can be driven by integrated circuits. A simple form of such a
picture screen consists of two parallel glass plates with a system of X and Y conductor paths (2 per mm) on their inner surfaces and a 10 μm to 20 μm thick liquid
crystal layer between them. The plates are arranged in such a way that the conductor strips are perpendicular to each other (Fig. 39a).

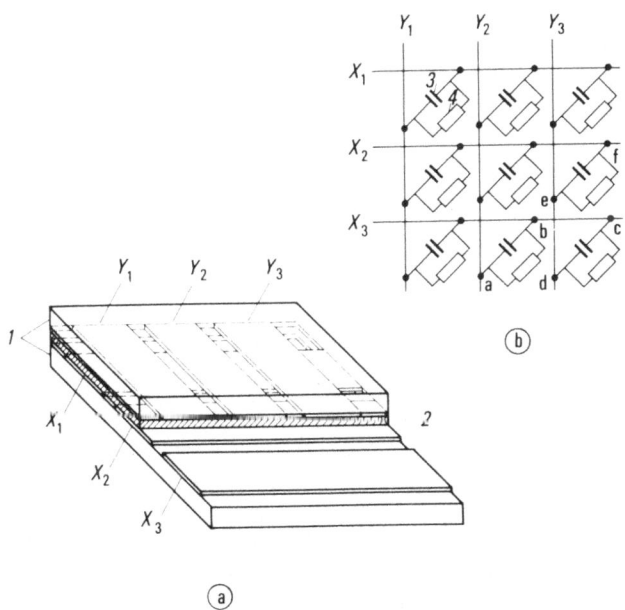

Figure 39. Matrix-type liquid crystal display.
(a) Schematic of the setup
(b) Circuit diagram

1 Glass plate	3 Capacitor
2 Liquid crystal layer	4 Resistor

X_{1-3} and Y_{1-3} Transparent conductor path

If a voltage U is applied to one of the X conductors, for example row X2, and one of the Y conductors, say column Y2, the lattice element corresponding to the intersection of X2 and Y2 sees the full voltage U. In the equivalent circuit diagram Fig. 39b, the liquid crystal is represented by a capacitor with a resistor in parallel. As a result of parasitic currents, voltages also appear at the other lattice elements, especially those in row X2 and column Y2, but they are always less than U/2. These elements are therefore also activated for instance to dynamic scattering, although much weaker.

This phenomenon observed in a liquid crystal matrix configuration and generally known as crosstalk is illustrated in Fig. 40. Parasitic currents can flow in the neighborhood of X2Y2 by galvanic coupling, for example as shown in Fig. 39b through abcdef. When driving the screen with short duration pulses of say 60 μs, not only galvanic coupling of the matrix points but also the capacitive coupling must be taken into account. The latter occurs in the same way as galvanic coupling [100].

Figure 40. Illustration of the crosstalk in a liquid crystal matrix configuration arising from parasitic currents.

By suitably designing the drive circuits, the voltages on all conductors not driven can be prevented from exceeding the value U/3, for example by applying the voltage U/3 to all X rows except X2 and 2U/3 to all Y columns excpet Y2. With a simple picture screen, crosstalk only fails to occur when the voltages U/2 resp. U/3 do not exceed the liquid crystal threshold voltage.

The turn-on time of presently available liquid crystals operating in the dynamic scattering mode is relatively long. According to Kochlman and Felici [101] the turn-on time of a nematic layer is given by the expression

$$t_{on} = A \cdot \frac{\eta}{\epsilon \left(\frac{U}{d}\right)^2},$$

where A is a constant, η the viscosity of the liquid crystal, ϵ the dielectric constant of the liquid crystal, U the voltage applied to the layer and d is the layer thickness.

Fig. 41 shows the turn-on time as a function of the voltage applied to the layer of a commercially available nematic liquid crystal with the layer thickness as parameter [69]. The diagram clearly illustrates that the turn-on time with dynamic scattering decreases sharply with increasing voltage. It also shows that the turn-on time for a liquid crystal with a threshold voltage of 8 V to 10 V and a layer thickness of 12 μm is about 10 ms for twice and about 1 ms for three times the threshold voltage. A turn-on time of 10 ms means that at least this time is required to write a picture line. At a picture frequency of 25 Hz this gives a picture screen with a maximum of 4 lines. A turn-on time of 1 ms with the same picture frequency results in a screen with a maximum of 40 lines.

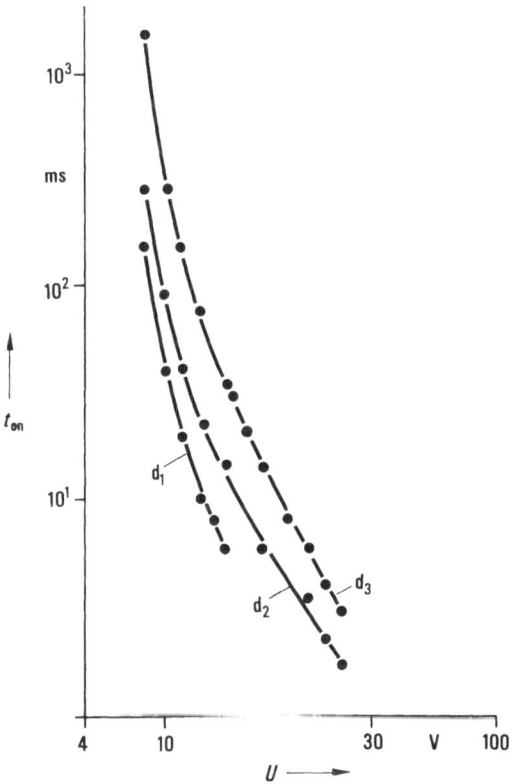

Figure 41. Turn-on time t_{on} of the dynamic scattering mode versus voltage U with the layer thickness d as parameter; d_1 = 6 μm, d_2 = 12 μm, d_3 = 25 μm (after reference [69]).

The threshold voltage of existing nematic liquid crystals is relatively low, and a matrix element excited to dynamic scattering even with twice or three times the threshold voltage still exhibits a relatively long turn-on time. This renders it impossible at present to construct picture screens having large information content without resorting to special "tricks" in the electronic drive circuits, additional temporary storage of the information or storing the information in the liquid crystal layer.

A number of techniques have been developed to produce matrix picture screens with a larger number of lines or points despite the inadequate threshold behavior.

These techniques can be divided into two basic groups [100]:

a) in the first group the picture screen contains no additional nonlinear elements in series with the picture elements of the liquid crystal screen; the maximum achievable number of lines for the screen is limited by the liquid crystal properties.

b) the second group of picture screens contains nonlinear switching elements in series with the liquid crystal elements. In this way universal picture screens can be built up. However, they are technologically very complex and presently not economically justifiable.

2.8.2. Matrix Arrays without Additional Devices in Series with a Liquid Crystal Element

2.8.2.1. Matrix Array Using Two-frequency Addressing

It was found [102] that the formation of Williams domains [103] and the dynamic scattering caused by d.c. or low frequency a.c. voltages can be suppressed by a superimposed a.c. voltage with an amplitude greater than that of the d.c. voltage and a frequency above the cutoff frequency f_c for domain formation (Fig. 42). Thus if a voltage $U_1 + U_2$ is applied to a matrix row, where U_1 is a d.c. or low frequency a.c. voltage greater than the threshold voltage U_{th} and U_2 is an a.c. voltage of suitable amplitude and frequency, and a voltage $-U_1 + U_2$ is applied to a column, scattering is only observed in the fully selected cell, which does not see the

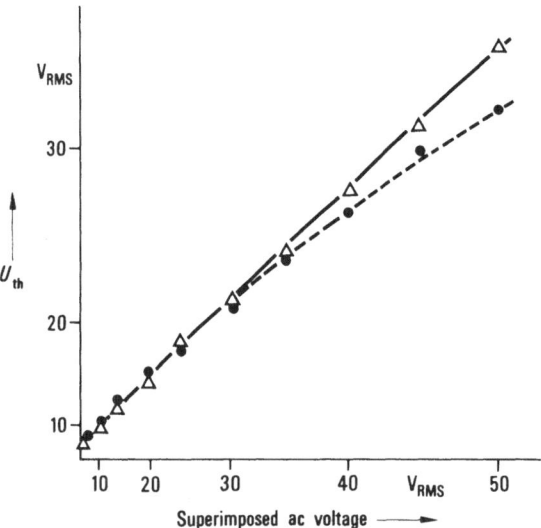

Figure 42. Dependence of the threshold voltage U_{th} for domain formation on the superimposed a.c. voltage (1,5 kHz), for d.c. (broken curve) and a.c. (50 Hz). (Quadratic scale for abscissa and ordinate). (After reference [102]).

suppressing signal and therefore scatters with an intensity corresponding to $2U_1$. A result obtained with a 3 × 7 matrix without crosstalking is shown in Fig. 43. In addition to the enhanced contrast, a turn-on time was observed at least three times faster than that under normal driving conditions. This means that in matrix-addressed liquid crystal light valve arrays, the number of rows can be increased by superimposing an a.c. electric field of sufficiently high frequency.

Figure 43. 3 × 7 liquid crystal matrix array without crosstalking using two frequency-addressing (after reference [102]).

The matrix addressing scheme is shown in Fig. 44. The voltage across the selected cells X_1Y_2, X_3Y_2 and X_4Y_2 is $2U_1$. No voltage is applied to the unselected cells X_2Y_1 and X_2Y_3. The remaining cells shown in Fig. 44 are partially selected and subjected to a combined signal of U_1 and $2U_2$. The amplitude and frequency of U_2 are so chosen that the electrooptic effect caused by U_1 alone can be suppressed. The matrix is addressed column by column. After each scan the controller reverses the polarity of the applied d.c. voltages U_1 to avoid undesirable electrochemical reactions in the liquid crystal layer.

2.8.2.2. Liquid Crystal Matrix Display by Electrically Controlled Birefringence

The homeotropic orientation of a nematic liquid crystal layer with negative dielectric anisotropy sandwiched between two electrodes can be deformed by an electric field. By increasing the field strength, the phase shift between the ordinary and extraordinary ray of incident light increases and results, if the cell is placed between crossed polarizers, in electrically controlled birefringence [86] (Section 2 6.1.).

This principle can also be extended to a liquid crystal matrix display with a larger number of independent points representing colored images [104]. In a matrix composed of 50 rows and 50 columns operated line-by-line, and with a picture frequency of 30 Hz, the on-time of each point is rather short:

$$t_{on} = \frac{1}{50 \times 30} \, s \approx 0.6 \, ms$$

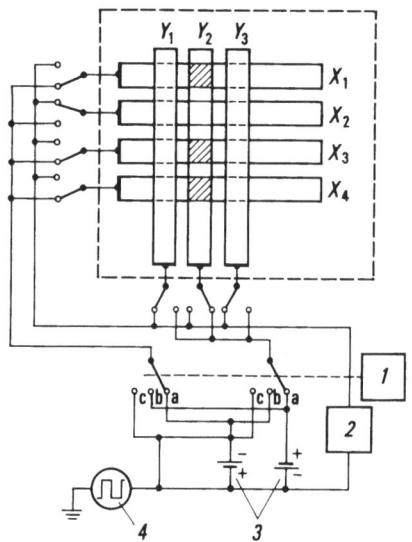

Figure 44. Matrix addressing scheme using two frequency addressing (after reference [102]).
X_{1-4} and Y_{1-3} Conductor path
1   Scan Controller
2   Phase shifter, $\Delta\gamma = 180°$
3   Batteries for voltage U_1
4   Oscillator for voltage U_2

This is too fast for nematics operated in dynamic scattering, but it is usable in electrically controlled birefringence (ECB). In ECB the nematic can respond to short duration pulses if the repetition frequency is high enough. The liquid crystal can summate a series of small excitations. If τ is the pulse duration and T the repetition period, the optical retardation does not depend on τ or T if these quantities are larger than the molecule relaxation time. The only important parameter is τ/T.

In a crossed electrode matrix the scheme $U_{th} - 3U_{th}$ can be used: the row pulses have an amplitude of $3U_{th}$, superimposed on a baseline level of U_{th} whereas column pulses extend from 0 to $2U_{th}$. That means 3×4.5 V = 13.5 V may be applied to an excited point, whereas the voltage applied to the other points is 4.5 V.

The threshold voltage is 4.5 V in d.c. operation. For an excited point the voltage is on during a small fraction of the time and the threshold increases. With τ/T = 1/50 and $U_{th} \approx 10$ V, the excess voltage above threshold is 13.5 V – 10 V = 3.5 V, which is too low to obtain a fast response; the response time is about 1s.

Higher voltages of 15 V to 18 V must be applied to the excited point to obtain satisfactory results. On the unexcited points there is a voltage of about 6 V, and hence a colored image on a background also colored and not dark is observed. With a filter a contrasted single color image on a dark background can be obtained. A time of 50 ms is necessary to change the image.

In order to make substantial progress and reach television standards, compounds with higher thresholds are required. As the time constant decreases with increasing voltage, it is necessary to apply voltages exceeding 20 V. With an elementary cell it is possible to achieve a response time as short as 10 μs with a 100 V pulse. When working with a crossed electrode matrix having a black background, no more than three times the d.c. threshold voltage can be applied. As U_{th} is a function of the dielectric anisotropy, nematics with a small negative — or positive — anisotropy must be found. With such products it will become possible to obtain a colored image on a dark background.

2.8.2.3. Matrix Display Based on the Texture Changes Cholesteric $\overset{\rightarrow}{\leftarrow}$ Nematic

This matrix display utilizes the properties of a mixture of a nematic liquid crystal that has positive anisotropy with about 10% cholesteric liquid crystal. With no electric field applied, the mixture behaves like a cholesteric material and scatters incident light, giving it a white appearance. But with an applied electric field (see also first part) there is a phase transformation to nematic homeotropic orientation and the material becomes transparent. If the electric field is removed, the liquid crystal reverts to the initial light scattering focal conic mode [105–107].

The characteristic of these texture changes is that a bias voltage that mantains an electric field below the threshold level increases the speed at which the liquid crystal transforms from the scattering to the transparent state, and decreases the speed at which it reverts from the transparent to the opaque state. The existence of a threshold level and the ability to slow the transformation back to the scattering state by applying a field below the threshold level makes it possible to use a convenient X–Y matrix arrangement [108] : in an experimental setup similar to other liquid crystal displays, 100 alpha-numeric characters with each character consisting of a 5 X 7 dot rectangle, in 4 lines each of 25 characters, were displayed. To scan the entire display takes 2.1 s. Even this relatively slow scanning speed produces no flicker. The contrast ratio for a reflective mode display was about 15:1.

2.8.2.4. Matrix Display Based on a Bistable Behavior of the Texture Changes Cholesteric $\overset{\rightarrow}{\leftarrow}$ Nematic

Cholesteric liquid crystal layers with positive dielectric anisotropy ($\Delta\epsilon > 0$) and with rigid perpendicular alignment of the molecules at the boundaries of the cell due to surface treatment exhibit a bistable behavior of the texture change from cholesteric to nematic and vice versa [109, 110]. The initial configuration of the liquid crystal molecules with no field applied to the layer is sketched in Fig. 45a. In the bulk there is a twisted structure with the helix axis being parallel to the plates, near the plates there is perpendicular alignment. In this state the texture is focalconic — the helix axis deviating more or less strongly from a direction parallel to the plates — causing diffuse scattering and depolarization of light.

When an electric field is applied perpendicular to the plates, the long molecular axes tend to align parallel to the field (purely dielectric alignment). This leads to the decrease of the twist with increasing field (Fig. 45b), resulting finally at a critical field $F_c\uparrow$

$$F_c\uparrow = \frac{\pi^2}{p}\frac{4\pi k_{22}}{\Delta\epsilon}$$

(p pitch, k_{22} twist elastic constant, $\Delta\epsilon$ dielectric anisotropy) in the nematic homeo-

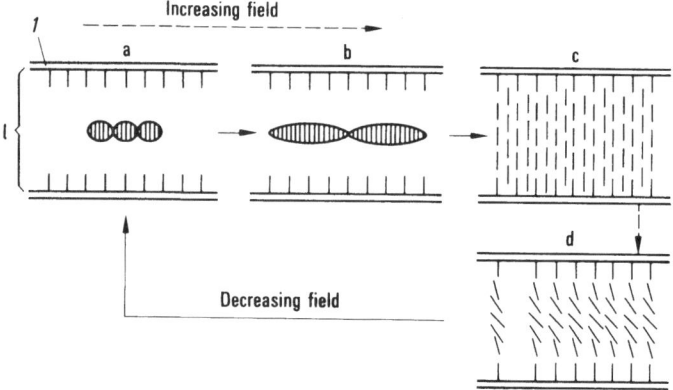

Figure 45. Scheme of the bistable behavior of the texture change cholesteric ⇄ nematic in a cholesteric liquid crystal layer with positive dielectric anisotropy if an electric field is applied. The length of the lines indicates the orientation of the molecular axis with respect to the plane of the drawing (after reference [110]).
(a) Initial focalconic configuration of the liquid crystal molecules with no field applied to the layer
(b) Decrease of the twist in the bulk with increasing field
(c) Homeotropic structure at a critical field $F_c\!\uparrow$
(d) Transition from homeotropic to focalconic mode with decreasing field.
1   Electrode plate,
l   Distance of the plates

tropic texture (Fig. 45c). The field-induced nematic texture is optically clear and polarized light propagating perpendicular to the plates is not influenced.

This texture transition from focal-conic to nematic is irreversible, provided there is perpendicular alignment at the cell boundaries. The helix structure forms differently with a decreasing electric field than with an increasing field, as shown in Fig. 45d. Also the critical field strength $F_c\!\downarrow$

$$F_c\!\downarrow \;=\; \frac{\pi}{p}\sqrt{\frac{4\pi}{\Delta\epsilon}\left(\frac{4k_{22}^{\,2}}{k_{33}}-k_{33}\left(\frac{p}{l}\right)^{2}\right)},$$

(k_{33} splay elastic constant, l distance of the plates) at which this transformation from the transparent back to the scattering mode occurs, differs from the critical field strength $F_c\!\uparrow$ at which the transition in the opposite occurs. For commercial liquid crystal mixtures $F_c\!\uparrow/F_c\!\downarrow$ = 2 to 3. The cell, being placed between crossed polarizers, exhibits a light transmission-voltage characteristic with a pronounced bistable behavior as illustrated in Fig. 46.

This effect can be utilized in a liquid crystal matrix display. In the initial state all matrix elements are switched to the nematic homeotropic (dark) state by switching all column electrodes to a high potential and all row electrodes to zero potential. Immediately after this operation the column electrodes are brought to a sustaining voltage U_s corresponding to a field strength between the two critical

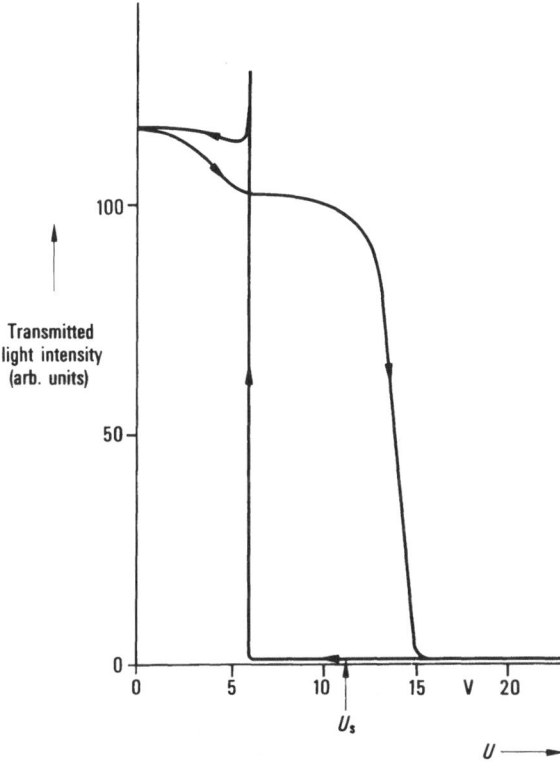

Figure 46. Transmitted light intensity versus voltage (quasistatic hysteresis curve) of a cholesteric liquid crystal layer with postive anisotropy; the cell being placed between crossed polarizers. Time per run: 3 min. (After reference [110]).

field strengths $F_c\uparrow$ and $F_c\downarrow$. This reduction in voltage to the sustaining voltage has no effect on the nematic state.

For writing, a matrix element is switched from dark to light, which occurs if the sustaining voltage on this element is switched off for a short time resulting in a transition from the nematic to the scattering focalconic state. This state is maintained, that means the information is stored until a voltage is applied which causes $F_c\uparrow$ to be exceeded.

In a representative liquid crystal mixture with a helix-pitch of about 2 μm, the transformation from focalconic to nematic homeotropic mode needs 20 ms at 80 V. The sustaining voltage U_s is about 10 V. The time required for the transition from the nematic to the focalconic mode is also about 20 ms.

The information is written a row at a time. For the representative mixture, about 20 ms are required to address one row. Because of the relatively long rise-time the effect described here may be applied in displays in which visual information has to be stored for a longer period and can be utilized, for instance, in alphanumeric information displays.

2.8.2.5. Matrix with Inherent Picture Storage

Cholesteric mixtures both with negative (Fig. 47) (see also first part of this book) and positive (Fig. 48) dielectric anisotropy that can be switched in both directions between two stable textures are suitable for use in inherent storage picture screens. Because of the long switching times of about 80 ms a larger storage picture screen cannot be realized with a cholesteric substance having negative dielectric anisotropy. On the other hand a X–Y matrix can be built up with mixtures having positive dielectric anisotropy [109].

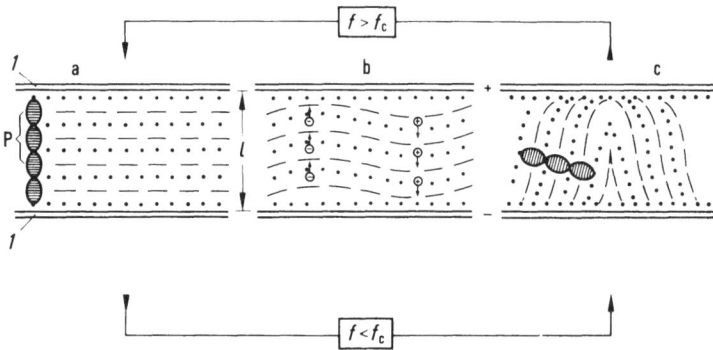

Figure 47. Storage effect in liquid crystals with negative dielectric anisotropy. Schematic of the texture changes under the influence of field strength and frequency (after reference [110]).
(a) Planar cholesteric initial texture of the layer; applied voltage: $V = 0$, p Pitch, 1 Electrode plate, l Distance of the plates, f Frequency, f_c Cutoff frequency;
(b) Periodically deformed texture of the layer; applied voltage: $V = V_{th}$.
(c) Focalconic texture of the layer; applied voltage: $V > V_{th}$.

If an electric field F is applied to a cholesteric liquid crystal layer with positive dielectric anisotropy, with $l/p < 4$ (where l is the layer thickness, p the helix pitch) with $k_{22}/k_{33} < \pi^2/4$, and the layer initially exhibiting a planar cholesteric texture (Fig. 48a), the field rotates the helix axis h perpendicular to the field and the layer assumes the light-scattering focalconic texture (Fig. 48b). When $h \perp F$ is maintained, the helix pitch increases with the field strength until at a field strength F_c (Section 2.8.2.4.) the liquid crystal changes to a nematic texture. When the field is removed the liquid crystal reverts rapidly, forming a twisted planar transition texture for a short time (Fig. 48d), to the original planar cholesteric texture (Fig. 48a).

The advantages of electrically-induced texture changes in cholesteric substances with positive dielectric anisotropy compared with the texture changes in cholesteric substances with negative dielectric anisotropy are:
a) it is a pure field effect.
b) for both writing and erasure d.c. voltages or d.c. voltage pulses can be used.
c) appreciably shorter write and erase pulses are possible. The write-in time is presently a few milliseconds, the erase time about 10 ms.

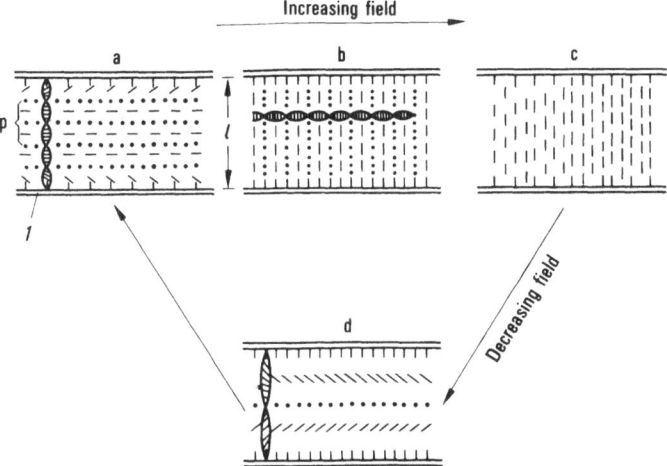

Figure 48. Storage effect in liquid crystals with positive dielectric anisotropy. Schematic of the texture changes under the influence of field strength (after reference [110]).
(a) Planar cholesteric initial texture of the layer
(b) Focalconic texture
(c) Nematic homeotropic texture
(d) Twisted planar transition texture
1   Electrode plate, p Pitch, l Distance of the plates

   The special property of cholesteric substances with positive dielectric anisotropy is that for mixtures with large $\Delta\epsilon$ the scattering focalconic texture is already converted to the transparent nematic homeotropic texture by a very short erase pulse of only approximately 50 μs. The remaining erase time is required for the folding back process to a planar texture. Thus the stored image can be erased rapidly.

2.8.2.6. Thermally Addressed Electrically Erased High Resolution Liquid Crystal Matrix Display

Soref's observations [111] that entire layers of cholesteric liquid crystal change from a scattering texture to the isotropic phase when heated temporarily can be made use of in designing a picture screen. A practical construction uses the efficient light scattering of thermally induced electrically erasable disordered regions in an otherwise well ordered clear cholesteric liquid crystal layer to form bright color-free black-and-white images with high contrast. In a typical system, the light scattering regions are thermally induced with an IR laser beam and are erased with an a.c. electric field [112].

   A typical thermally induced light valve consists (Fig. 49a) of a thin layer (14 μm) of cholesteric liquid crystal sandwiched between two substrates coated with visibly transparent IR-absorbing electrodes such as $In_{2-x}Sn_xO_{3-y}$. An X–Y deflected intensity-modulated focused IR laser beam is absorbed by the electrodes, thereby locally heating the liquid crystal from its well-ordered transparent planar cholesteric state to the transparent isotropic phase as shown in Fig. 49b. Cooling occurs almost instantaneously after the laser beam is deflected to a new location, leaving the addressed area with disorder frozen into the light scattering focalconic cholesteric phase (Fig. 49c). The disordered regions scatter light strongly. Thus any arbitrary pattern of scattering centers can be written into the light valve by appropriately scanning a focused IR laser beam across it. Erasure is accomplished by applying an a.c. field of typically 40 V_{rms} at 1.5 kHz between the transparent electrodes. The erasure time is less than 1 s and decreases with increasing electric field.

Usually a cholesteric liquid crystal is chosen with negative dielectric anisotropy to facilitate erasure, a pitch much shorter than the cell thickness to facilitate storage, and a pitch longer than the wavelength of the viewing light as measured in the liquid crystal to minimize Bragg scattering in the erased state.

The image quality obtained by thermal addressing is quite satisfactory. Lines with 20 μm- to 50 μm-wide scattering regions are obtained when 3 mW to 10 mW of power is absorbed from a laser beam moved at speeds of up to 10 cm/s across the liquid crystal cell. These thermally addressed liquid crystal light valves thus feature high resolution and require only moderate addressing power, especially when preheated to a few centigrades below the cholesteric-isotropic phase transition temperature.

Potential applications for these thermally addressed liquid crystal light valves include recording, storing and displaying handwriting, facsimile and high resolution graphic information on a front or back illuminated projection screen. Light valves with 2000 × 2000 resolvable spots and resolutions of 50 lines/mm are readily

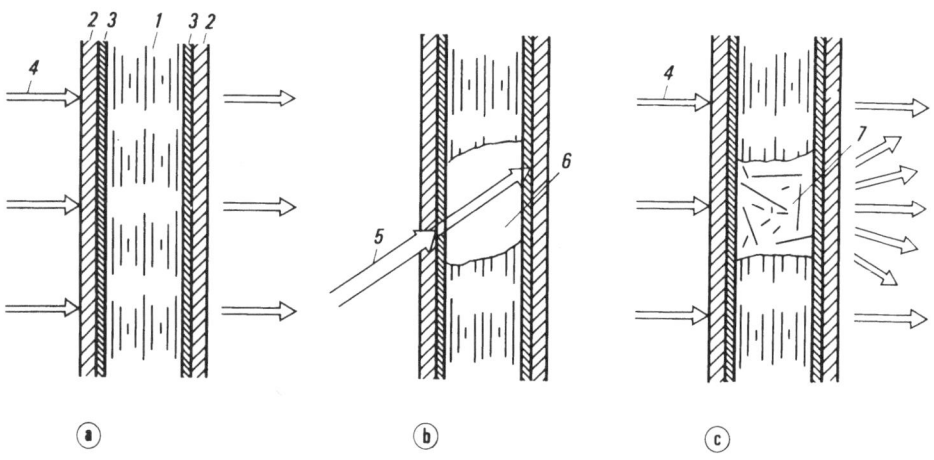

Figure 49. Construction and operation of a thermally addressed liquid crystal matrix display (after reference [112]).
(a) Off-state
(b) Writing
(c) Display in the addressed and storing mode
1   Planar cholesteric liquid crystal layer   5   Infrared laser beam
2   Substrate   6   Addressed area in the isotropic phase
3   $In_{2-x}Sn_xO_{3-y}$ Electrode   7   Addressed area in the scattering focalconic
4   Viewing Light   mode

achievable. Addressing speeds in excess of 10^4 picture elements/s have been demonstrated by using a 1.06 μm laser beam that generates a few milliwatts of heating power. Since the thermal writing devices require neither charge carriers in the liquid crystal nor d.c. or low frequency a.c. fields to be applied to the liquid crystal, lifetimes should be considerably longer than those of any other light valve structures previously proposed.

2.8.3. Matrix Displays with Additional Nonlinear Devices in Series with a Liquid Crystal Element

There are basic properties that an individual element of a universal matrix display must exhibit [113]:

a) the element must possess a threshold voltage.

b) the elements should store the information during the time taken to address all other elements of the display.

c) in order to avoid smearing, the elements must turn on and turn off within the period of one frame.

d) if a gray scale is desired, for example for television, the contrast must be continuously variable.

Since a liquid crystal cell does not possess an adequate threshold for large matrix selection, the threshold must be provided by an additional device or devices at each element.

2.8.3.1. Matrix Displays with Additional Diodes

In a simple matrix scheme the threshold is provided by an additional diode at each element [113] as shown in Fig. 50. The slow decay in brightness of the liquid crystal is used to provide frame storage. All diodes in this scheme are biased off except those coincidentally selected by row and column signals.

Assuming that the diode has low forward resistance and the source impedance of U_s (t) is also low, the cell voltage U_c (t) will easily reach U_s in a rather short time, e.g. in 60 μs for television application (Fig. 51a–c). Neglecting diode leakage, i.e. assuming the reverse resistance of the diode to be large compared to the resistance of the liquid crystal cell, U_c (t) decays with the dielectric relaxation time constant of the liquid crystal material which may typically be about 10 ms or 30 ms (Fig. 51c).

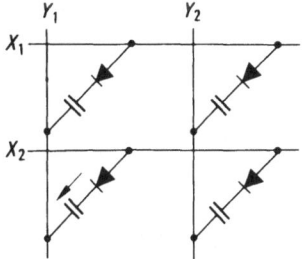

Figure 50. Liquid crystal matrix array with additional diodes.

Since the liquid crystal material cannot respond instantaneously to the applied
voltage, the voltage across the cell has decreased to say $(1/3)U_c$ at the time the cell
begins to scatter light. Since U_c (t) continues to decay, the brightness response also
starts to decay soon after it has started to rise as shown in Fig. 51d.

Though this simple addressing scheme does work, it has the following disadvantages:
a) low average brightness,
b) large addressing voltages,
c) severe diode requirements,
d) dependence of the brightness response on the relaxation time.

Therefore Lechner et al. [113] proposed a more suitable addressing scheme
(Fig. 52). In this scheme two diodes and a capacitor are associated with each cell
in the matrix. The purpose of the capacitor is appreciably to extend the relaxation
of the liquid crystal cell. Several important advantages are gained over the former
scheme by doing this. First, since the applied addressing voltage is maintained

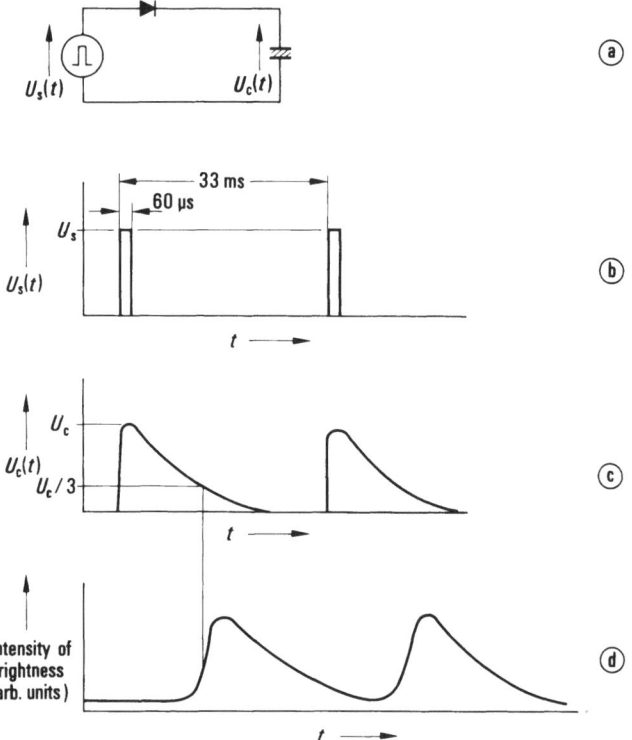

Figure 51. Operation of a liquid crystal display element in series with an additional diode (after
reference [113]).
(a) Equivalent circuit for an element in the matrix; U_s Source voltage, U_c Cell voltage
(b) Addressing signal
(c) Voltage across cell
(d) Brightness response

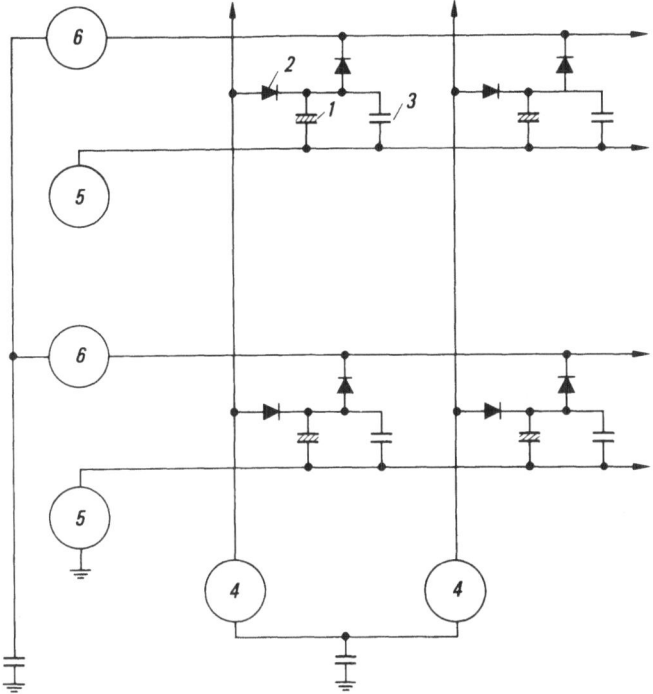

Figure 52. Double-Diode-Capacitor addressed liquid crystal matrix display (after reference [113]).

1	Liquid crystal cell	4	Positive column pulse generator
2	Diode	5	Negative row pulse generator
3	Capacitor	6	Negative reset pulse generator

across the liquid crystal cell for the entire frame period, the required addressing voltage to achieve saturated brightness from the liquid crystal is substantially reduced. Second, the liquid crystal cell voltage waveform is now unaffected by the somewhat unpredictable dielectric relaxation time of the liquid crystal material. Third, since the brightness is maintained approximately constant for the entire frame period, the average brightness is higher than that obtained without the capacitor. Fourth, the requirements on the diodes need not be as severe as for the former addressing scheme. The disadvantage of this scheme is the more complex circuitry.

2.8.3.2. Thin-film-transistor Matrix-addressed Transparent TV Panel Based on Twisted Nematic Liquid Crystal Light Valves

A display with a twisted nematic liquid crystal layer represents an ideal light valve, which, between parallel linear polarizers, a) is dark in the off-state, b) is bright in the on-state with less than 10 V, c) has usable grey scale, d) has adequate response time and contrast ratio, e) has long life and can be made with normal window glass.

By combining twisted nematic liquid crystal light valves with thin film transistors, Fischer [114] and Fischer et al. [115] showed that a transparent TV panel can be

built. The illumination by white light through a mosaic colour film filter also makes
color pictures feasible. The panel consists, from top to bottom (Fig. 53), of a light
diffuser foil to permit wide angle viewing, a linear polarizing foil and the twisted
nematic cell. The transparent base plate of the cell carries the thin film transistor
matrix array containing the multiple transparent back electrode pads for the twisted
nematic liquid crystal elementary light valves. On its protruding edges this base
plate carries the peripheral scanning circuitry (scanners, video gates, storage
register), either co-deposited on the same panel, or on coextensive sub-panels (Fig.
54) connected to the main panel by silk screened silver-epoxy contacts. The cell is
followed by a linear polarizer foil identical to the above and a color film sheet with
mosaic-arrayed red, green and blue filter rectangles side by side to form square
triplets separated by a black raster, in exact registration with the thin film transistor
matrix above. A plastic honeycomb louver sheet transmits only collimated or near-
collimated light. For illumination a flat white light source such as a white fluorescent lamp with suitable reflectors, an a.c. electroluminescent panel, or an edge-
illuminated glass plate with a raster of frosted spots or a bonded-on lenticular
plastic foil are used. Top and base plates of the panel are spaced by peripheral
Mylar shims and evaporated SiO dots on the thin film transistor matrix, and are

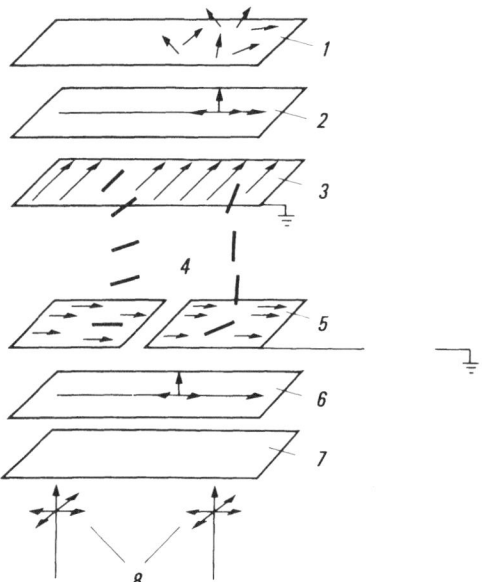

Figure 53. Thin-film-transistor matrix-addressed transparent TV panel based on twisted
nematic liquid crystal light valves (after reference [115]).

1 Frosted foil   5 Base plate with orienting surfactant
2 Linear polarizer foil   and with the film matrix array
3 Transparent front electrode with   6 Linear polarizer foil
   orienting surfactant   7 Color filter
4 Nonactivated and activated twisted nematic 8 Light source
   layer

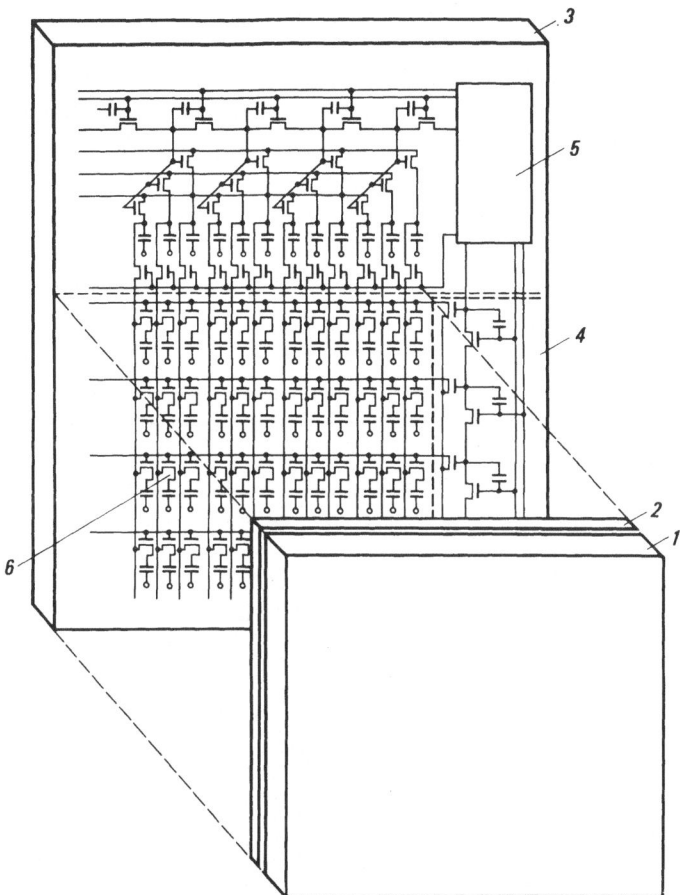

Figure 54. Thin film transistor matrix display with peripheral scanning circuitry (after reference [115]).

1	Transparent front electrode	4	Subpanel with row scanners
2	Liquid crystal layer	5	Subpanel with column scanners
3	Transparent back electrode	6	Subpanel with thin film transistors

sealed at the edges with resin. — A 120 × 120 matrix on a 6″ × 6″ glass plate has been made.

For addressing the incoming amplitude-modulated video is channeled by the horizontal, shift-register-scanned video gates into the capacitors of the storage register. When filled, this storage register is discharged simultaneously into all columns during the extended flyback time of 20 μs. The slow vertical scanner has selected one row in which it turns on all thin film gates for 60 μs. The matrix thin film transistors in this row admit the video pulses to the elemental light valve capacitors. When the next row is addressed, the previously addressed thin film transistors cut off and trap the charge in the capacitors. There it stays until the next addressing occurs, one frame later. The polarity of the video is reversed each frame period, so that the twisted nematic liquid crystal is driven by 15 Hz a.c. square waves of varying amplitude. This improves the life of the liquid crystal and of the thin film transistors. The thin film transistors are only used as on-off switches. Their off-resistance has to be higher than 10^{10} Ω.

In the meantime an integrated 14,000 picture element 2,3 dm^2 (36 in^2) flat screen display panel has been constructed by Brody [116] et al. by combination of thin film transistor and nematic liquid crystal technology.

2.8.3.3. Liquid Crystal Matrix Displays Using an Additional Ferroelectric Layer for Providing the Necessary Electrical Threshold and for Storing Video Signals

For an a.c. voltage, a ferroelectric layer represents a capacity which varies with the amplitude of the applied voltage. This can be seen from the familiar hysteresis loops, of which two specific cases are shown in Fig. 55: The large loop is traversed for an amplitude V, the small one for V/2. The slopes of the broken lines are proportional to the effective dielectric constant, that is, to the capacity of the ferroelectric layer. If the ferroelectric ceramic possesses a sufficiently square hysteresis loop, below the saturation region the capacity increases very strongly with the amplitude of the applied voltage.

A double layer, consisting of a ferroelectric ceramic and a liquid crystal layer, therefore represents for a.c. voltage pulses the series connection of a nonlinear and linear capacity. By suitably dimensioning the capacities of both layers, the voltage across the liquid crystal layer can be increased much stronger than linear with the total voltage applied across the double layer, as shown in Fig. 56. By this method

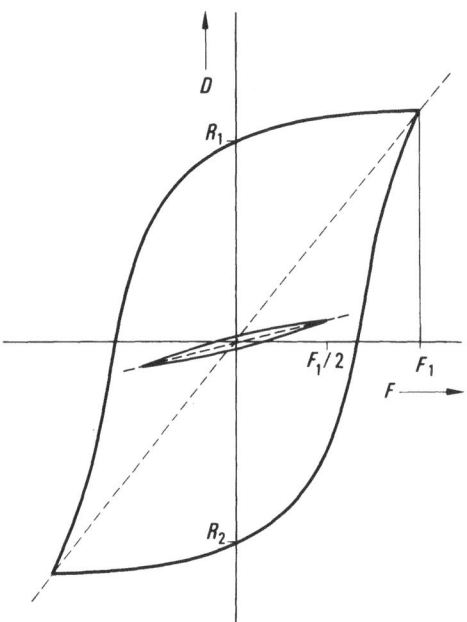

Figure 55. Hysteresis loops of a ferroelectric ceramic at a field strength of F_1 or $F_1/2$; dielectric polarization D versus F. R_1 upper, R_2 lower remanence point (after reference [117]).

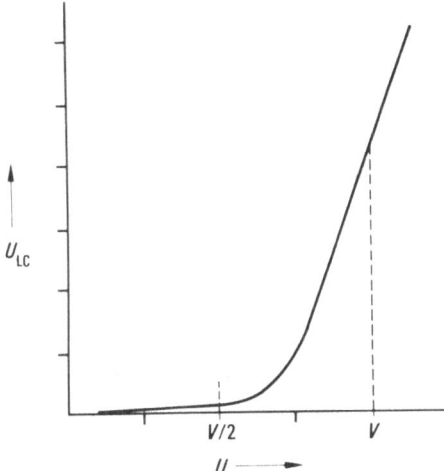

Figure 56. Voltage amplitude U_{LC} across the liquid crystal layer as a function of the total voltage U across the double layer, consisting of a ferroelectric ceramic and a liquid crystal layer (after reference [117]).

the necessary electrical threshold voltage of the display elements in a liquid crystal matrix can be obtained [117, 118]. An example is demonstrated in Fig. 57, which shows a single display element excited by matrix addressing. The double layer is provided on one side with the X-, and on the other side with the Y-conductors. Because of the great differences in the dielectric constants of the ferroelectric ceramic and the liquid crystal, it is necessary amongst other things to match the capacity by using different widths of the X- and Y-conductor strips. This can be realized for example in a setup shown in Fig. 58a and Fig. 58b.

Figure 57. Laboratory model of a liquid crystal matrix display, each liquid crystal element in series with a ferroelectric ceramic element. One element (ca. 0,1' × 0,1') is excited.

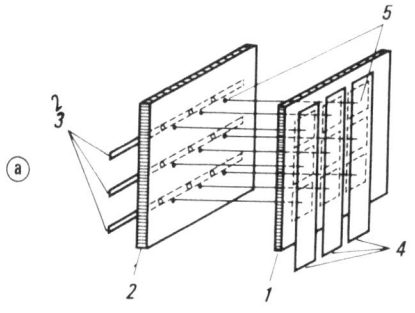

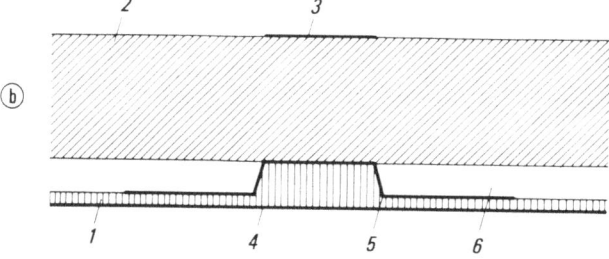

Figure 58. Integrated setup of a liquid crystal matrix display with an additional ferroelectric layer (after reference [117]).
(a) Fundamental array
(b) Cross section of an element

1 Liquid crystal layer	4 Transparent X-conductor strips
2 Ferroelectric layer	5 Reflecting intermediate electrode
3 X-Conductor strips	6 Isolating layer

By addressing the ferroelectric ceramic elements of the matrix display must be polarized by the addressing pulses alternately in positive and negative field directions. This can be done in practice, for example, by writing one frame with positive pulses and then applying across all elements at the same time one negative pulse, which is so short that the liquid crystal will not be excited but all ceramic elements will be polarized inversely. Then again, the next frame is written with positive pulses etc.

In the initial state, let all ceramic elements be in the upper remanence point R_1 of the hysteresis loop (Fig. 55). By applying negative d.c. voltage pulses across the conductor matrix (X and Y conductors), the individual elements are polarized inversely, for example, into the lower remanence state R_2. The polarization process takes about 1 μs. The image information is thus first converted into a polarization pattern stored in the ferroelectric ceramic. This polarization pattern can now be converted at the same moment into a visible image on the liquid crystal film by applying a single positive readout pulse of about 1 ms across all X-conductors together, and the common front electrode. This readout pulse causes all ferroelectric elements to return to the initial state R_1. During this flipping back of the ferroelectric domains, polarization charges corresponding to the magnitude of the preceding polarization will flow to the respective liquid crystal capacitor elements, thus exciting the liquid crystal. After that, the next frame can be written into the ceramic, etc.

The degree of brightness of the display elements is controlled by the amplitude of the write pulses. The voltage pulses in the experiments had an amplitude of about 100 V. The rise-times of the nematic liquid crystal used were several milliseconds and the decay times were of the order of 100 milliseconds. Since the resistivity of the ferroelectric ceramic is substantially greater than that of the liquid crystal, the ferroelectric layer can also be provided with a bias voltage.

The ceramic materials used consisted of lead zirconate–lead titanate mixtures. These materials have relatively square hysteresis loops and can be prepared in large thin plates. The thickness of the ceramic layer was of the order of 100 μm, that of the liquid crystal about 10 μm. With a thickness of about 50 μm, these ceramic mixtures have a sufficient transparency, so that a matrix which also operates in transmitted light can be realized.

Because of the ability of the ferroelectric ceramic to store video signals as polarization states, liquid crystal picture screens with 24 frames/s analogous to the normal TV screen can be displayed in spite of the long rise times of the liquid crystals.

2.9. Application of Nematic Liquid Crystals in Coherence Optics

2.9.1. General Considerations

From the optical standpoint the outstanding properties of dynamic scattering are that it can be electrically controlled in two dimensions and the active motion of the scattering centers. By utilizing the property that dynamic scattering can be electrically controlled, it is possible to build electrooptic devices with transmission characteristics that can be controlled in two dimensions and as a function of time. Such devices can lead to new real time techniques in ultrasonic and microwave holography and in coherent optics data processing.

An important assumption with these applications is that the coherent light can be amplitude-modulated, whereby only the unscattered part of the light passing through the liquid crystal layer is used and its amplitude controlled by the applied voltage. This appears possible on two conditions:
a) it must be possible to completely separate the scattered and unscattered light,
b) the coherence of the unscattered part must not be affected by the scattering process.

Some intensity modulation can be achieved from light scattering alone where the liquid crystal is employed as a light valve, because the scattered light is no longer available for displaying the image. A further property of light scattering liquid crystal layers is that the scattered light is not coherent with the unscattered light and can therefore be completely separated for holographic displays.

R. Bartolino et al. [119] have investigated the spatial coherence of light passing through a liquid crystal initially homogeneously or homeotropically oriented as a function of the applied external d.c. field and of the molecule orientation to the direction of polarization of the incident light.

The results of their study show that liquid crystals may be used to change the degree of spatial coherence of an initially coherent light beam and that the dimensions of the coherence region are determined by the applied electric field strength.

The initial state of polarization of the incident beam determines in some cases the ability of the liquid crystal to change the beam coherence.

Finally it is noted that the results can provide information about the way molecular motion occurs.

Some practical applications of these properties can be envisaged. For example, the possibility of fringe shift as a function of the field strength F can be utilized for designing phase modulators or filters. When the beam coherence is reduced, it changes statistically in the sense that the size of the coherence area is decreased by increasing F, but the location of the coherence regions in the sample is somehow random. On the other hand the coherence properties of light scattered by the liquid crystal can be exploited. Then liquid crystals can be used in nearly the same way as a rotating ground glass.

2.9.2. Data Input Matrix (Page Composer) for Hologram Storage

Strong dynamic scattered laser light cannot be registered holographically because of its incoherence. This allows the input data for holographic data storage to be controlled electrically. The problem here is to "switch off" arbitrary combinations from a larger number of light sources, i.e. to prevent their holographic registration. The basic experiment described in the following demonstrates that this problem can be solved with liquid crystal cells.

In the experiment a raster of 5 × 5 points is recorded holographically, whereby arbitrary combinations of these points can be switched off (Fig. 59a). The point raster is formed by the focal points of lenses arranged in a raster and illuminated by collimated light. A liquid crystal matrix display is located in front of the lens raster. Each element of the liquid crystal cell is allocated to one lens.

The path of the rays during recording is shown in Fig. 59b. The laser beam is divided by a prism filter. The reference beam is deflected by a mirror through a microscope lens onto the photographic plates. A telescope system spreads the other part of the laser beam and directs it through the liquid crystal cell and lens raster onto the hologram.

The configuration described can be used to take various storage holograms. This is equivalent to recording various point combinations. Those light points representing binary "0" are suppressed by the dynamic scattering of the driven liquid crystal element. Light points representing binary "1" are not suppressed.

Fig. 60 shows the holographic reconstruction and a conventional picture from the same viewing angle. The hologram contains binary "0" 14 times. The conventional non-holographic picture clearly illustrates the difference between those lenses illuminated by collimated light and those illuminated by diffuse light.

An experimental model of a large optical memory has been built at RCA [120] which, it is claimed, will replace the entire hierarchy of magnetic tape, disc, drum and core memories during the 1980's because of its exceptionally high capacity and the capability of electronic access with this device.

The following briefly outlines the design and application of this novel high-

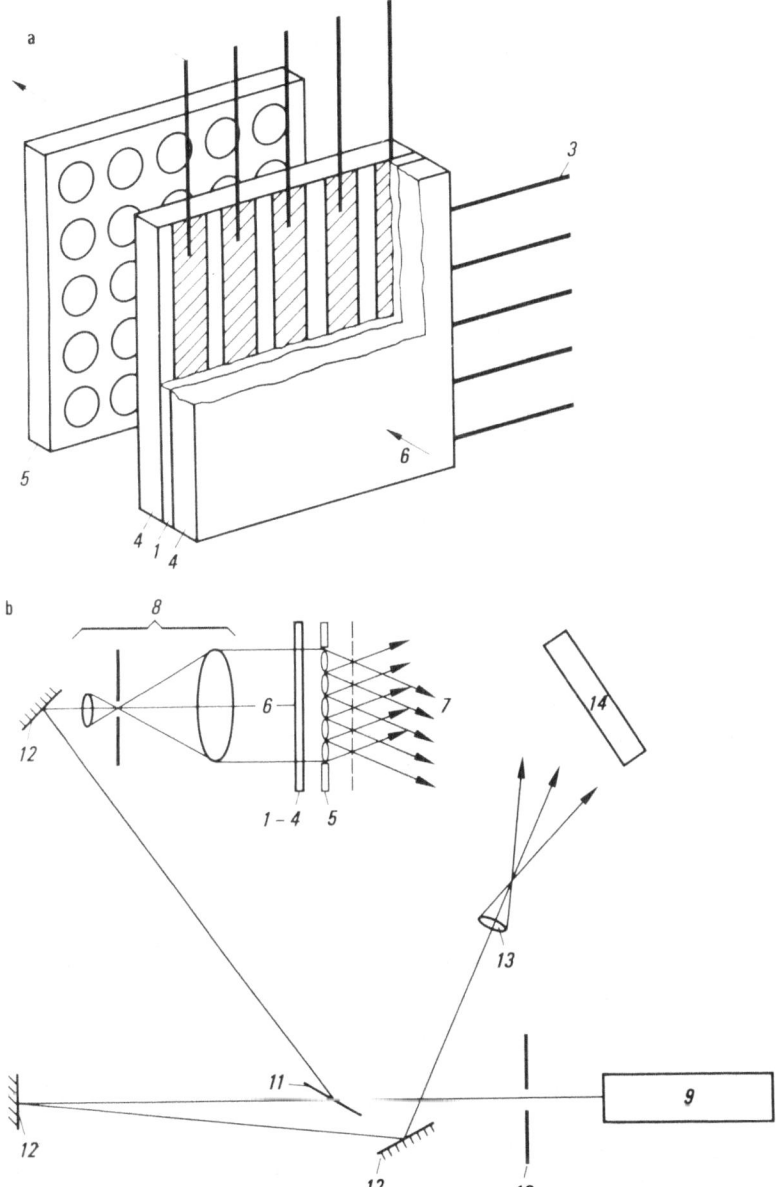

Figure 59. Data input matrix with liquid crystal layer for hologram storage.
(a) Matrix array
(b) Path of rays during recording

1   Liquid crystal layer
2,3 Conductor strips
4   Glass plate
5   Lens raster
6   Laser illumination
7   Object wave
8   Telescope system

9 He-Ne-laser
10 Camera shutter
11 Prism filter
12 Mirror
13 Microscope lens
14 Photographic plate, hologram

Figure 60. Holographic reconstruction (left) and conventional picture (right) of a lens arrangement combined with a liquid crystal matrix display.

capacity memory. A beam generated by a pulsed ruby laser passes through two electronically controlled electroacoustic crystals and is deflected depending on the frequency of the sound waves to which the crystals are subjected. One of these crystals deflects the beam horizontally, the other vertically. Hence there are 32×32 different positions spatially for the beam leaving the second crystal.

The beam then falls on 1024 different holograms on a hololens surface. The hololens splits the beam into two parts; the first passes unhindered through the lens whereas the second is diffracted and falls on a square, measuring 76 mm $\times$ 76 mm. This square consists of 1024 tiny liquid crystal elements which can be electronically made to reflect or transmit light. With the help of these cells digital information from the minute surfaces is gated into the laser beam when they are either dark, with the liquid crystal cell operating in the reflective mode, or light when the cells are operating in the transmissive mode. These surfaces correspond to binary "1" or "0" and represent the information to be written into the memory.

The coded or modulated laser beam is now directed to one of 1024 partial surfaces on a manganese-bismuth film where it combines again with the first directly incident part of the beam. This produces a magnetic hologram on the selected partial surface. The magnetic equivalent of the optical interference pattern forming on the film is generated by the laser beams converging on it. In this way the information pattern presented by the liquid crystals is stored holographically.

The information is read out by a second laser beam deflected onto the hologram of the manganese-bismuth film but not onto the liquid crystals. Erasure is similar to writing except that neither of the beams is modulated or coded by the liquid crystals. The point on which the two laser beams meet is therefore magnetically neutralized. The film suffers no damage, and the write, read and erase processes can be repeated at will.

2.9.3. Granulation-free Picture Screens

Virtually all objects around us scatter light to some degree. When a light scattering surface is illuminated by coherent light, it does not appear uniformly light to the observer but exhibits a more or less granular structure with completely homogeneous illumination. This phenomenon is known as granulation or speckle patterns [121]. It is based on the fact that elementary spherical waves are scattered after the Huygens principle from the scattering centers of the surface, which are all mutually coherent, interfere with each other and hence form a stationary interference pattern. As the phase of their reflection varies statistically from scattering center to scattering center, the interference field is a statistical distribution of light and dark. This is why the surface appears speckled. When images are projected with coherent light, granulation is annoying. The larger the image detail the more displeasing granulation becomes.

The small moved regions of the nematic liquid crystal layer act as a light scattering center with an electric field applied. In daylight the layer looks like a very fine-grain ground-glass plate and can be used as such or as a projection screen. It turns out that unlike conventional ground-glass plates or screens no granulation can be observed. Fig. 61 shows such an experimental model. Fig. 61a is a test picture projected onto a conventional ground-glass plate, in Fig. 61b the ground-glass plate has been replaced by a liquid crystal layer. The unmistakable difference in the test pictures is explained by the fact that the scattering liquid crystal layer destroys the coherence of the light. As a result the interference field of the scattered elementary waves (speckle pattern) changes its structure so fast that the human eye can no longer follow because of its inertia.

The effect described here has so far only been achieved with mechanically movable, for example rotating, ground-glass plates and screens. Movable parts are, however, generally undesirable in equipment, and they can also not be manufactured to any exact size and precision.

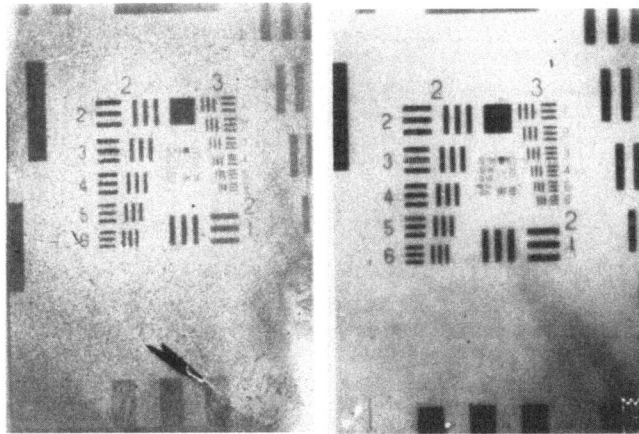

Figure 61. Granulation free liquid crystal picture screen. Comparison of a test picture projected onto a conventional ground glass plate (a) and a liquid crystal layer in the dynamic scattering mode (b).

Acknowledgments. The author of this section of the book is indebted to Dr. Hölzler and Professor Dr. Welker for their approval to carry through this work at the Siemens Research Laboratories in the frame of his regular duties.

Special thanks are due to Mr. Krüger for numerous and valuable suggestions and discussions as well as for his careful and critical reading of the manuscript. Finally, I want to thank Mr. Peach for preparing the English version of this book section.

References

1. Selawry, O., Holland, I., Selawry, H.: Mol. Cryst. 1, 445 (1966).
2. Gros, C., Gautherie, M., Bourjat, P., Archer, F.: Ann. Radiol. 13, 333 (1970).
3. Selawry, O. S., Selawry, H. S., Holland, J. F.: Mol. Cryst. 1, 495 (1966).
4. Peterson, E. N., Dixon, S. D.: Obstetrics and Gynecology 37, 438 (1971).
5. Horn, J., Steinsträßer, R.: Anwendungsmöglichkeiten cholesterinischer flüssiger Kristalle in Medizin und Technik, Referat, Techn. Akademie Esslingen (22.7.74).
6. Koehnlein, E., Dietrich, F.: Med. Tribune No. 21, 26.5.72.
7. Brown, G. H. et al.: Liquid crystals AD 725949, Techn. Rep. AFML-TR-71-20 April 1971.
8. Carroll, P. L.: Cholesteric liquid crystals, their technology and applications, 22 Grays Inn Road, London WC1X 8HR, England: Ovum Ltd. (1973).
9. Ennulat, R. D., Fergason, J. L.: Mol. Cryst. Liq. Cryst. 13, 149 (1971).
10. Klein, E. J.: Astronautics and Aeronautics 5, 1120 (1972).
11. Klein, E. J.: Astronautics and Aeronautics No. 70 AIAA Paper July 1968.
12. Magura, K.: Nachrichtentechn. Z. 9, 440 (1970).
13. Iizuka, K.: Proc. IEEE 58, 288 (1970).
14. Sproat, W. H., Cohen, S. E.: Mater. Eval. 28, 73 (1970).
15. Lukianoff, G. V.: Mol. Cryst. Liq. Cryst. 8, 389 (1967).
16. Jones, C. H., Fergason, J. L., Arars, J. A.: AD 629550 (Westinghouse).
17. Fergason, J. L.: Appl. Optics 9, 1729 (1968).
18. Steinsträßer, R.: Anwendungsmöglichkeiten cholesterinischer flüssiger Kristalle in Medizin und Technik, Referat, Technische Akademie Esslingen (16.2.73).
19. The National Cash Register Co., Dayton Ohio USA: Auslegeschrift 1 698 092, Anmeldetag 24.2.1968.
20. Hansen, J. R., Schneeberger, R. J.: IEEE Trans. ED-15, 11, 896 (1968).
21. Haas, W., Adams, J.: Appl. Optics 7, 1203 (1968).
22. Heilmeier, G. H.: Sci. Amer. 222, 100 (1970).
23. Goodman, L.: J. Vac. Sci. Technol. 10, 5, 840 (1973).
24. Soref, R.: Proc. S.I.D. 13, 95 (1972).
25. Heilmeier, G., Zanoni, L., Barton, L.: IEEE Trans. ED-17, No. 1, 22 (1970).
26. Heilmeier, G., Goldmacher, J.: Proc. IEEE 57, 34 (1969).
27. Scheffer, T.: J. Appl. Physics 11, 4799 (1973).
28. Kmetz, A.: IEEE Trans. ED 20, 11, 954 (1973).
29. Creagh, L.: Proc. IEEE 61, 814 (1973).
30. Nishimura, Y., Yamamoto, T., Inagaki, T., Sasaki, H.: New Method. Fujitsu Scientific and Technical Journal 2, 135 (1966).
31. Fraser, D. B.: Thin Solid Films 13, 407 (1972), Proc. IEEE 61, 1013 (1973).
32. Vossen, J., Poliniak, E.: Thin Solid Films 13, 281 (1972).
33. Dreyer, J. F.: Third Internat. Liquid Crystal Conf. Berlin, Session 1.1, 24 (1970).
34. Greubel, W., Krüger, H., Wolff, U.: dtsch. Offenlegungsschrift 2163 606.
35. Kahn, F.: Appl. Phys. Letters 22, 386 (1973).
36. Uchida, T., Watanabe, H., Wada, M.: Japanese J. Appl. Physics 11, 1559 (1972).
37. Proust, J., Ter-Minassian-Saraga, L., Guyon, E.: Solid State Comm. 11, 1227 (1972).
38. Krüger, H., Mahlein, H., Rauscher, W.: dtsch. Offenlegungsschrift 2330 909 (1974).
39. Kahn, F., Taylor, G., Schonhorst, H.: Proc. IEEE 61, 823 (1973).
40. Berreman, D.: Phys. Rev. Lett. 28, 1683 (1972).

41. Grabmaier, J. et al.: Presented at the 4th Int. Liquid Crystal Conf. Kent, Ohio, Paper 103, (August 1972).
42. Wolff, U., Greubel, W., Krüger, H.: Mol. Cryst. Liq. Cryst., 23, 187 (1973).
43. Janning, J.: Appl. Phys. Lett. 21 (1972).
44. Dixon, G., Brody, T., Hester, W.: Appl. Phys. Lett. 24, 47 (1974).
45. Creagh, L., Kmetz, A.: Presented at the 4th Int. Liquid Crystal Conf. Kent, Ohio, Paper 20, (August 1972).
46. Plueddeman, E.: J. Adhes. 2, 184 (1970).
47. Kashnow, R.: Rev. Sci. Instr. 43, 1837 (1972).
48. Bayer, H.: Dtsch. Offenlegungsschrift 2159 165 (1972).
49. Greubel, W., Krüger, H., Wolff, U.: Dtsch. Offenlegungsschrift 2251 998.
50. Kobayashi, S., Shimojo, T., Kasano, K., Tsunda, I.: S.I.D. Symposium (1972) Paper 5.4.
51. Soref, R.: S.I.D. Internat. Sympos. Digest Techn. Papers New York USA (1973).
52. Stein, C.: S.I.D. Intern. Sympos. Digest Techn. Papers, New York USA, 38 (1973).
53. Soref, R.: Appl. Optics 9, 1323 (1970).
54. Beard, T., Bleha, W., Wong, S.: Appl. Phys. Lett. 22, 90 (1973).
55. Margerum, J., Nimoy, J., Wong, S.: Appl. Phys. Letters 17, 51 (1970).
56. Jacobson, A. D., Beard, T. D., Bleha, W. P., Margerum, J. D., Wong, S. Y.: Pattern Recognition, Pergamon Press New York 5, 13 (1973).
57. Margerum, J., Wong, S.: SPSE Symposium on Unconventional Photographic Systems, Washington, DC (Oct. 1971).
58. Flannery, J.: IEEE Trans. Ed-20, 11, 941 (1973).
59. Schadt, M., Helfrich, W.: Appl. Phys. Lett. 18, 127 (1971).
60. Bonne, U., Gummings, J.: IEEE Trans. ED-20, 11, 962 (1973).
61. Grabmaier, J., Knauer, R., Krüger, H.: ETZ-B 25, 626 (1973).
62. Geffcken, W., Krüger, H.: Konstrukteur 5, 5 (1974).
63. Brown, C. H., Shaw, W. G.: Chem. Rev. 57, 1049 (1957).
64. Dave, J. S., Patel, P. R.: Mol. Cryst. 2, 103, 115 (1966).
65. Gray, G. W.: Molecular Structure and the Properties of Liquid Crystals. New York: Academic Press (1962).
66. Gray, G. W.: Mol. Cryst. 5, 333 (1966); Mol. Cryst. Liq. Cryst. 7, 127 (1969).
67. de Jeu, W. H., v. d. Veen, J., Goossens, W. J. A.: Solid State Commun. 12, 405 (1973).
68. Kast, W.: Angew. Chem. 67, 592 (1955).
69. Wiegand, H.: Z. Naturforsch. 6b, 240 (1951).
70. Creagh, L.: Proc. IEEE 61, 814 (1973).
71. Steinsträßer, R., Pohl, L.: Angew. Chem. 85, 706 (1973).
72. Kast, W., Börnstein, L.: 6. Aufl. Ba II/2a, 266 Berlin: Springer-Verlag 1960.
73. Rondeau, R. E., Berwick, M. A., Steppel, R. N., Serré, M. P.: J. Am. Chem. Soc. 94, 1096 (1972).
74. de Jeu, W. H., v. d. Veen, J.: Philips Res. Rep. 27, 172 (1972).
75. Knaak, L. E., Rosenberg, H. M., Serré, M. P.: Mol. Cryst. Liq. Cryst. 17, 171 (1972).
76. Kelker, H., Scheuerle, B.: Angew. Chem. 81, 903 (1969), Angew. Chem. (Intern. Edit.) 8, 884 (1969).
77. Steinsträßer, R., Pohl, L.: Z. Naturforsch. 26b, 87 (1971).
78. Steinsträßer, R.: Z. Naturforsch. 27b, 774 (1972).
79. Steinsträßer, R., Pohl, L.: Z. Naturforsch. 26b, 577 (1971).
80. Steinsträßer, R., Pohl, L.: Tetrahedron Lett. 1921 (1971).
81. Boller, A., Scherrer, H., Schadt, M., Wild, P.: Proc. IEEE Lett. 60, 1002 (1972).
82. Eliott, G.: Chem. Brit. 9, 5, 213 (1973).
83. Sussmann, A.: Appl. Phys. Lett. 21, 126 (1972).
84. Voinov, M., Dunnett, J. S.: J. Electrochem. Soc. 120, 7, 922 (1973).
85. Briere, G., Herino, R., Mondon, F.: Mol. Cryst. Liq. Cryst. 19, 157 (1972).
86. Denat, A., Gosse, B., Gosse, J.: J. Chim. phys. 70, 319 (1973).
87. Lomax, A., Hirasawa, R., Bard, A. J.: J. Electrochem. Soc. 119, 679 (1972).
88. Schiekel, M. F., Fahrenschon, K.: Appl. Phys. Lett. 19, 10, 391 (1971).

89. Assouline, G., Hareng, M., Leiba, E.: Electronics Letters 7, 699 (1971).
90. R. A. Soref, M. J. Rafuse.: J. Appl. Phys. 43, 5, 2029 (1972).
91. Kahn, F. J.: Appl. Phys. Lett. 20, 5, 199 (1972).
92. Greubel, W., Wolff, U.: Appl. Phys. Lett. 19, 213 (1971).
93. Heilmeier, G. H., Zanoni, L. A.: Appl. Phys. Lett. 13, 3, 91 (1968).
94. Greubel, W., Krüger, H., Wolff, U.: Siemens-Z. 46, 11 862 (1972).
95. Kobayashi, S., Takeuchi, F.: SID Internat. Symposium Digest of Technical Papers New York, USA, 15–17.5.73, 40 (1973).
96. Kmetz, A. R., IEEE Trans. ED-20 No. 11, 954 (1973).
97. Hareng, M., Assouline, G., Leiba, E.: 4th Int. Liq. Cryst. Conf. Kent, Ohio, Paper 84 (1972).
98. Stein, C. R., Kashnow, R. A.: SID Symp. Dig. 64 (1972).
99. Becker, J. H., Wysocki, J. J., Dir, G. A , Madrid, R. W., Adams, J. E., Haas, W. E., Leder, L. B., Bechlowitz, B., Saera, F. D.: SID Symp. Dig. (1971).
100. Grabmaier, J. G., Krüger, H. H.: VDI-Z. 115, No. 8, 629 (1973).
101. Kochlmans, H., van Boxtel, A. M.: Phys. Lett. 32A, 32 (1969). N. Felici, Rev. Gen. Elec. 78, 717 (1969).
102. Wild, P. J., Nehring, J.; Appl. Phys. Lett. 19, 335 (1971).
103. Stein, C. R., Kashnov, R. A.: Appl. Phys. Lett. 19, 343 (1971).
104. Hareng, M., Assouline, G., Leiba, E.: 4th Int. Liq. Cryst. Conf. Kent, Ohio, Paper 84 (1972).
105. Wysocki, J. J., Adams, J., Haas, W.: Phys. Rev. Letters 20, No. 19, 1024 (1968).
106. Baessler, H., Labes, M. M.: J. Chem. Phys. 51, 5397 (1969).
107. Ohtsuka, T., Tsukamoto, M.: Japan J. Appl. Phys. 12, 22 (1973).
108. Electronics: Oct. 23 (1972).
109. Greubel, W.: Appl. Phys. Lett. 25, 5 (1974).
110. Krüger, H., Walter, K.: Ber. Bunsen-Ges. 78, 912 (1974).
111. Soref, R. A.: J. Appl. Phys. 41, 3022 (1970).
112. Melchior, H., Kahn, F., Maydan, D., Fraser, D.: Appl. Phys. Lett. 21 No. 8, 392 (1972).
113. Lechner, B. J., Marlowe, F. J., Nester, E. O., Tults, J.: Proc. IEEE 59, 1566 (1971).
114. Fisher, A. G.: IEEE Conf. Display Devices New York (Oct. 1972).
115. Fischer, A. G., Brody, T. P., Escott, W. S.: IEEE Conf. Conf. Display Devices New York (Oct. 1972).
116. Brody, T. P., Asars, J. A., Dixon, D. G.: IEEE Trans. ED-20, No. 11, 995 (1973).
117. Grabmaier, J. G., Greubel, W. F., Krüger, H. H.: Mol. Cryst. Liq. Cryst. 15, 95 (1971).
118. Tannas, L. E., York, P. K.: SID Internat. Sympos. Digest of Techn. Papers New York, USA, 15 (1973).
119 Bartolino, R., Bertolotti, M., Scudieri, F., Sette, D.: Appl. Optics 12, No. 12, 2917 (1973).
120. Stewart, W. C., Mezrich, R. S., Cosentina, L. S., Nagle, E. M., Wendt, S. S., Lohman, R. D.: RCA Rev. 34, 3 (1973).
121. Kiemle, H., Röss, D.: Einführung in die Technik der Holographie, 102, Akademische Verlagsgesellschaft Frankfurt am Main (1969).

Subject index

Chemical Compounds

H. Rickert
Einführung in die Elektrochemie fester Stoffe
64 Abbildungen. XV, 223 Seiten. 1973
Gebunden DM 46,–; US $ 18.90
ISBN 3-540-06266-1

Die Elektrochemie fester Stoffe ist in den letzten Jahren weit
über das rein wissenschaftliche Interesse hinaus gewachsen
und hat an Bedeutung und technischen Anwendungsmöglichkeiten gewonnen. Wenige Stichworte genügen, um das anzu-
deuten: Galvanische Ketten mit festen Elektrolyten, Fehlord-
nungen in festen Stoffen, Transportvorgänge, Festkörperreak-
tionen, Halbleiter. Chemiker, Physiker, Metallurgen und Werk-
stoffwissenschaftler sowie vor allem auch Studenten finden
hier Grundlagen und Beispiele einer modernen Forschungs-
richtung verständlich dargestellt.

Electrons in Fluids
The Nature of Metal-Ammonia Solutions
Editors: J. Jortner and N. R. Kestner
271 figures. 59 tables. XII, 493 pages. 1973
Cloth DM 120,–; US $ 49.20
ISBN 3-540-06310-2

This full and up-to-date account of the chemical and physical
properties of electrons in polar, nonpolar, and dense fluids includes contributions from both theoretical and experimental
chemists and physicists, thus clearly indicating the interdis-
ciplinary nature of this field.

E. Fitzer, W. Fritz
Technische Chemie
Eine Einführung in die Chemische Reaktionstechnik
Hochschultext
150 Abbildungen. 36 Tabellen, 31 Rechenbeispiele.
XIII, 552 Seiten. 1975
DM 44,–, US $ 18.10
ISBN 3-540-06787-6

Das Buch wendet sich vorwiegend an Studenten. Es gibt zu-
nächst einen Überblick über die wirtschaftlichen Grundlagen
der chemischen Produktion sowie das Produktspektrum der
chemischen Industrie und umreißt dann die Aufgaben der
Chemischen Reaktionstechnik. Nach einer Einführung in die
physikalisch-chemischen Grundlagen werden die Methoden
zur Berechnung chemischer Reaktoren für homogene, hetero-
gene und Polymerreaktionen behandelt.

Lyotropic Liquid Crystals
18 figures. 3 tables. IV, 85 pages. 1975
(NMR Basic Principles and Progress, Volume 9)
Cloth DM 38,–; US $ 15.60
ISBN 3-540-07303-5

Lyotropic liquid crystals acquire their anisotropic properties
from the mixing of two or more components. One component
is amphiphilic and contains a polar head group, the second
component is usually water. Lyotropic liquid crystals occur
abundantly in nature, particularly in all living systems. The
most familiar example of a lyotropic liquid crystal is soap in
water. During the last few years considerable progress has
been made toward understanding the experimental and theoretical aspects of the method reviewed in this book.
(462 references).

Springer-Verlag
Berlin
Heidelberg
New York

Preisänderungen vorbehalten
Prices are subject to change without notice

Structure and Bonding

Editors: J.D. Dunitz, P. Hemmerich, R.H. Holm, J.A. Ibers, C.K. Jørgensen, J.B. Neilands, D. Reinen, R.J.P. Williams

Volume 20
Biochemistry
57 figures. IV, 167 pages. 1974
Cloth DM 66,–; US $ 27.10
ISBN 3-540-07053-2
Contents: A.S. Mildvan, C.M. Grisham: The Role of Divalent Cations in the Mechanism of Enzyme Catalyzed Phosphoryl and Nucleotidyl Transfer Reactions. – H.P.C. Hogenkamp, G.N. Sando: The Enzymatic Reduction of Ribonucleotides. – W.T. Oosterhuis: The Electronic State of Iron in Some Natural Iron Compounds: Determination by Mössbauer and ESR Spectroscopy. – A. Trautwein: Mössbauer-Spectroscopy on Heme Proteins.

Volume 21
Recent Impact of Physics on Inorganic Chemistry
62 figures. IV, 144 pages. 1975
Cloth DM 58,–; US $ 23.80
ISBN 3-540-07109-1
Contents: B.C. Tofield: The Study of Covalency by Magnetic Neutron Scattering. – B. Frick: Superheavy Elements.

Volume 22
Rare Earths
36 figures. IV, 175 pages. 1975
Cloth DM 69,–; US $ 28.30
ISBN 3-540-07268-3
Contents: E. Nieboer: The Lanthanide Ions as Structural Probes in Biological and Model Systems. – C.K. Jørgensen: Partly Filled Shells Constituting Anti-bonding Orbitals with Higher Ionization Energy than their Bonding Counterparts. – R.D. Peacock: The Intensities of Lanthanide f ↔ f Transitions. – R. Reisfeld: Radiative and Non-Radiative Transitions of Rare-Earth Ions in Glasses.

Volume 23
Biochemistry
50 figures. IV, 193 pages. 1975
Cloth DM 74,–; US $ 30.40
ISBN 3-540-07332-9
Contents: J.A. Fee: Copper Proteins – Systems Containing the "Blue" Copper Center. – M.F. Dunn: Mechanisms of Zinc Ion Catalysis in Small Molecules and Enzymes. – W. Schneider: Kinetics and Mechanism of Metalloporphyrin Formation. – M. Orchin, D.M. Bollinger: Hydrogen-Deuterium Exchange in Aromatic Compounds.

Volume 24
Photoelectron Spectrometry
40 figures. 21 tables. IV, 169 pages. 1975
Cloth DM 68,–; US $ 27.90
ISBN 3-540-07364-7
Contents: C.K. Jørgensen: Photo-electron Spectra of Non-metallic Solids and Consequences for Quantum Chemistry. – P.A. Cox: Fractional Parentage Methods for Ionisation of Open Shells of d and f Electrons. – R.E. Watson, M.L. Perlman: X-Ray Photoelectron Spectroscopy. Application to Metals and Alloys. – A.M. Bradshaw, L.S. Cederbaum, W. Domcke: Ultraviolet Photoelectron Spectroscopy of Gases Adsorbed on Metal surface.

Springer-Verlag
Berlin
Heidelberg
New York